Modern Printed-Circuit Antennas

Edited by Hussain Al-Rizzo

Published in London, United Kingdom

IntechOpen

Supporting open minds since 2005

Modern Printed-Circuit Antennas
http://dx.doi.org/10.5772/intechopen.77513
Edited by Hussain Al-Rizzo

Contributors
Yu Hongbiao, Pai-Yen Chen, Chung-Tse Michael Wu, Haipeng Li, Guang-Ming Wang, Haisheng Hou, Tong Cai, Xiangjun Gao, Wenlong Guo, Amer Tawfeeq Abed, Mahmood Jawad Abu-AlShaeer, Aqeel Mahmood Jawad, Jafar Ramadhan Mohammed, Yasir Al-Yasir, Abdulkareem Abdullah, Raed. Alhameed, Hasanain A. H. Al-Behadili, Baha A. Sawadi, Ahmed M. Abdulkhaleq, Sulaiman Tariq, Hussain Al-Rizzo, Ayman Isaac, Samer Yahya, Ashish Singh, Durga Prasad, Krishnaanand Shet, Nikolay Atanasov, Gabriela Atanasova, Blagovest Atanasov, Karam Mudhafar Younus, Naser Ojaroudi Parchin

First published in London, United Kingdom, 2020 by IntechOpen
IntechOpen is the global imprint of INTECHOPEN LIMITED, registered in England and Wales, registration number: 11086078, 5 Princes Gate Court, London, SW7 2QJ, United Kingdom
Printed in Croatia

British Library Cataloguing-in-Publication Data
A catalogue record for this book is available from the British Library

Additional hard and PDF copies can be obtained from orders@intechopen.com

Modern Printed-Circuit Antennas
Edited by Hussain Al-Rizzo
p. cm.
Print ISBN 978-1-83880-857-0
Online ISBN 978-1-83880-858-7
eBook (PDF) ISBN 978-1-83880-859-4

For EU product safety concerns:
IN TECH d.o.o., Prolaz Marije Krucifikse Kozulić 3, 51000 Rijeka, Croatia, info@intechopen.com or visit our website at intechopen.com.

Meet the editor

Hussain Al-Rizzo received his Ph.D. in Electrical and Computer Engineering from the University of New Brunswick, Fredericton, NB, Canada. In 2000 he joined the Systems Engineering Department, University Arkansas Little Rock where he is currently a Professor of Telecommunication Systems Engineering. He has published over 200 papers in peer-reviewed journals and conference proceedings, 6 book chapters, and 4 patents. His research areas include implantable antennas and wireless systems, smart antennas, massive MIMO systems, electromagnetic wave scattering by complex objects, design, modeling and testing of high-power microwave applicators, precipitation effects on terrestrial and satellite frequency re-use communication systems, field operation of NAVSTAR GPS receivers, data processing, and accuracy assessment, and the effects of the ionosphere, troposphere and multipath on code and carrier-beat phase GPS observations.

Contents

Preface

Recent developments in wireless communications such as 4G and the upcoming 5G cellular system, MIMO and massive MIMO, implantable and wearable systems, satellite, and radar systems, impose stringent design specifications and constraints. This book aims to present the latest developments in modern printed-circuit antennas written by experts in the field. This book consists of nine chapters. It is ideal for senior undergraduate students, graduate students, and engineers/researchers working in the field of antenna design.

Chapter One introduces a new concept to reduce mutual coupling among closely-spaced antenna elements of a MIMO array. This concept results in a significant reduction in the complexity of traditional approaches such as metamaterials, defected ground plane structures, parasitic elements, matching and decoupling networks using a planar metallic ring resonator printed on one face of an ungrounded substrate that surrounds a two-element vertical monopole antenna array. It is demonstrated both by simulations and measurements that the mutual coupling is reduced by at least 20 dB, maintaining the impedance bandwidth over which S_{11} is less than -10 dB, and reducing the envelope correlation coefficient to below 0.001. The two vertical monopoles are operating at 2.4 GHz and separated by 8 mm (λ_o /16), where λ_o is the free-space wavelength at 2.45 GHz.

In Chapter Two, recent research advances on beamforming and spatial multiplexing techniques using reconfigurable metamaterials (MTMs) and metasurfaces are reviewed. This chapter describes basic principles of transmission line-based metamaterials and planar metasurfaces, followed by their active versions that enable novel smart antennas with beam steering and beamshaping functions. Detailed descriptions of their practical realizations and the integration with circuits and the radio-frequency (RF) frontend are also discussed. The latest metasurface-based beamforming techniques are explained and compared for their uses in the RF-to-millimeter-wave range in terms of cost, reconfigurability, system integratability and radiation properties.

In Chapter Three a new ultrathin broadband reflected metasurface is proposed for applications involving high-gain planar antennas. A multilayer, multifunctional transmitted metasurface is next introduced to simultaneously enhance the gain and transform linear polarization to circular polarization of a patch antenna. This kind of high-gain antenna eliminates the feed-block effects of the reflected ones.

Chapter Four discusses how traditional size limitations can be overcome using a fractal geometry antenna. The shape is repeated into a limited size such that the total length of the antenna is increased to match, for example, half of the wavelength of the corresponding desired frequency. Many fractal geometries, e.g., the tree, Koch, Minkowski, and Hilbert fractals, are common. This chapter describes the details of designing, simulations, and experimental measurements of fractal antennas. Based on dimensional geometry in terms of desired frequency bands, the characteristics of each iteration are investigated in order to improve the antenna design process. The surface current distribution is analyzed to enhance the circular polarization radiation and axial ratio bandwidth. Simulation and experimental results are compared. Two types of fractal antennas are reported, the first has a new structure configured via a five-stage process. The second has a low profile, wherein the configuration of the antenna was based on three iterations.

Chapter Five introduces several planar array designs based on the use of a small number of the active elements located at the center of the planar array surrounded by another layer of the uniformly distributed parasitic elements. The parasitic elements are used to modify the radiation pattern of the central active elements. The overall radiation pattern of the resulting planar array with a small number of active elements is comparable to that of the fully active array elements with a smaller sidelobe level at the cost of a relatively wider beamwidth and lower directivity. Nevertheless, the use of only a small number of the active elements provides a very simple feeding network that reduces the cost and the complexity of the array. The proposed array can be effectively and efficiently used in applications that require wider antenna beams.

Reconfigurable beam steering using a circular disc microstrip patch antenna with a ring slot is presented in Chapter Six. The dimension of the antenna is $5.4 \times 5.4\ mm^2$ which is printed on a 0.504 mm thick, RT5870 substrate with relative permittivity 2.3 and loss tangent 0.0012. The antenna operates in the 60 GHz 5G frequency band. Two NMOS switches are utilized to generate three different beam patterns. Activating each switch individually results in a 70° shift in the main beam direction with constant frequency characteristics. The power gain is 3.9–4.8 dB in the three states of switch configurations. Simulated results in terms of return loss, peak gains, and radiation pattern are presented and reveal good performance for 5G applications.

In Chapter Seven, an ultra-wide band angular ring antenna that operated from 2.9 to 13.1 GHz is presented. The antenna has a nearly omni-directional radiation pattern for the E and H-plane at 3.5, 5.8, and 8.5 GHz. The theoretical analysis of the proposed antenna has been analyzed using circuit theory analysis. Measured, simulated, and analytical results compare well for reflection coefficient and radiation patterns. The proposed antenna is suitable for Wi-Fi, WiMax, digital communication system, satellite communication, and 5G applications.

Noise figure and noise power are analyzed and deduced in theory in Chapter Eight for multi-port networks in active phased array radar. The mathematical expressions of output noise power and noise figure of a network are provided under various conditions. Accordingly, this provides a basis of theories for a multi-port network and radar receiver system design, the test method of array noise figure. Two application examples are given to verify the accuracy of the formulas. Making use of these formulas, the designer can calculate the dynamic range of the radar receiver system, and the designer can also constitute a measure scheme of the array noise figure for active phased array radar.

Chapter Nine provides a brief overview of the types of wearable antennas possessing high body-antenna isolation. The main parameters and characteristics of wearable antennas and their design requirements are discussed. Next, procedures (passive and active) to test the performance of wearable antennas are presented. The electromagnetic properties of the commercially available textiles used as antenna substrates are investigated and summarized. This is followed by a more detailed examination of their effects on the performance of wearable antennas with high body-antenna isolation. A trade-off between substrate electromagnetic properties and resonant frequency, bandwidth, radiation efficiency, and maximum gain is presented. Finally, a case study is presented with detailed analyses and investigations of the low-profile all-textile wearable antennas with high body-antenna isolation and low SAR. Their interaction with a semisolid homogeneous human body phantom is discussed. The simulations and experimental results from different (in free-space and on-body) scenarios are presented.

I would like to gratefully acknowledge the Chair of the Systems Engineering Department, Dr. Ibrahim Nisnaci and Dr. Lawrence Whitman, Dean of the of Donaghey College of Science, Technology, Engineering, and Mathematics, University of Arkansas Little Rock for the encouragement they have offered throughout this project. Thanks are due to Dajana Pemac and Rebekah Pribetic, Author Service Managers, IntechOpen for the expert assistance offered.

Dr. Hussain Al-Rizzo
Professor of Electrical and Computer Engineering,
Department of Systems Engineering,
Donaghey College of Science, Technology, Engineering, and Mathematics,
University of Arkansas Little Rock,
Little Rock, AR, USA

Chapter 1

Decoupled and Descattered Monopole MIMO Antenna Array with Orthogonal Radiation Patterns

Hussain Al-Rizzo, Ayman A. Isaac, Sulaiman Z. Tariq and Samer Yahya

Abstract

This chapter introduces a novel design concept to reduce mutual coupling among closely-spaced antenna elements of a MIMO array. This design concept significantly reduces the complexity of traditional/existing design approaches such as metamaterials, defected ground plane structures, soft electromagnetic surfaces, parasitic elements, matching and decoupling networks using a simple, yet a novel design alternative. The approach is based on a planar single decoupling element, consisting of a rectangular metallic ring resonator printed on one face of an ungrounded substrate. The decoupling structure surrounds a two-element vertical monopole antenna array fed by a coplanar waveguide structure. The design is shown both by simulations and measurements to reduce the mutual coupling by at least 20 dB, maintain the impedance bandwidth over which S_{11}, is less than −10 dB, and reduce the envelope correlation coefficient to below 0.001. The boresight of the far-field radiation patterns of the two vertical monopole wire antennas operating at 2.4 GHz and separated by 8 mm ($\lambda_0/16$), where λ_0 is the free-space wavelength at 2.45 GHz, is shown to be orthogonal and inclined by 45° with respect to the horizontal (azimuthal) plane while maintaining the shape of the isolated single antenna element.

Keywords: 4G/5G LTE/LTE-A, cellular communications, coplanar waveguide fed vertical monopole, decoupled antenna array, descattered antenna array, GPS, microstrip fed vertical monopole, MIMO, radar, RFID, Wi-Fi, WiMAX

1. Introduction

Contemporary wireless systems including, but not limited to, 4G/5G LTE/LTE-A, radar, RFID, Wi-Fi, WiMAX, GPS, geolocation, biomedical imaging, and remote sensing dictate the use of miniaturized MIMO antenna arrays on mobile terminals. They can also be permanently installed on fixed structures for increased gain, which will improve link reliability and quality of service, increase communication range, and increase battery life through a variety of diversity schemes [1, 2], and/or increase data rate/throughput through MIMO spatial multiplexing schemes [3].

Moreover, these structures can provide esthetic, miniaturized wireless consumer devices. Recent trends toward miniaturized, esthetically appealing, battery efficient handheld wireless devices, and green wireless systems require antenna arrays that should be implemented within a restricted physical space.

In this chapter, we report a new approach of how to reduce the complexity of prior existing designs for mutual coupling reduction such as periodic metamaterial/metasurface constructs [4–8], defected ground plane structures [9, 10], soft/hard electromagnetic surfaces [11, 12], parasitic elements [13], matching and decoupling networks [14–16], and neutralization lines [17]. Our proposed approach can still maintain, and in certain circumstances exceed the performance metrics of traditional approaches well-known in previous art, [18–20] using much simpler, cost-effective, and novel design alternatives.

Numerous techniques were proposed for the mutual coupling reduction among the elements of MIMO antenna arrays, the most notable and highly embraced is the use of metamaterials, which subsequently turned into an unique topic of its own used in various disciplines including acoustics, RF, optics, laser, and nanotechnology [21]. A defected ground plane consisting of a slitted pattern is disclosed in [9] to suppress mutual coupling between two monopole antennas separated by 0.093 λ_0, where λ_0 is the free-space wavelength at 2.53 GHz. The two monopoles share a 40 × 25 mm ground plane and are separated by 11 mm. Each slit is 11 × 1 mm. A periodic array of three slits was used to achieve a mutual coupling better than −20 dB. However, no information on the behavior of far-field radiation patterns of the array was provided. A simple yet elegant decoupling approach was presented in [22] using a parasitic element to decouple two monopole antennas operating at 900 MHz. The parasitic element is situated at the center between the two driven elements on a 330 × 250 mm FR4 substrate, 1.5 mm in thickness. The substrate has a permittivity of 4.4 and a loss tangent of 0.02. The radiation behavior shows an end-fire pattern. The measured pattern correlation is close to 0.02. This method achieved coupling coefficient, S_{12} close to −40 dB over the bandwidth in which S_{11} is less than −10 dB. The method requires careful attention to the choice of the distance between the parasitic element and the driven elements for a given separation between the driven elements. A drawback of this technique is the undesired radiation from the parasitic element, which may degrade the decoupled radiation pattern.

A complicated approach for mutual coupling suppression between two monopole antennas operating at 2.49 GHz and separated by 0.09 λ_0, where λ_0 is the free-space wavelength at 2.49 GHz, with a length of 32.5 mm over a ground plane containing four slits, 1 × 11 mm in size and two defected wall structures situated at the edge of the outermost slits are proposed in [20]. The defected wall structure consists of a metallic plane with C-shaped cuts mimicking split ring resonators. The spacing between the slots in the defected wall is 0.09 λ_0. The structure achieved a measured isolation of about 44 dB. No information has been provided on the shape of the far-field radiation patterns. One notable approach for the isolation of two high-profile monopole antennas laid on a common copper ground plane (1.25 λ_0 × 1.25 λ_0), involves the use of highly negative-permeability, broadside coupled split-ring resonators etched on both sides of a dielectric substrate. In this approach, the two monopoles are separated by $\lambda_0/8$ (30.25 mm), where λ_0 is the free-space wavelength at 1.24 GHz [18]. A 3-D stack employing 10 arrays is aligned vertically using plastic support in the space between the two monopoles, where each array consists of four broadside coupled split-ring resonators.

It should be noted, however, that these artificial magnetic inclusions obviously suffer from extremely narrow resonance bandwidth. The design methodology is based on creating artificial negative permeability, which consequently presents extremely high attenuation to the near-field that exists in the region between the two monopoles for both propagating and evanescent fields. The authors claim a reduction of at least 20 dB in S_{12} but over an extremely small bandwidth. The structure is neither convenient nor rigid for miniaturized handheld devices since the decoupling structure occupies a significant volume in the space between the two monopoles. Also, the fabrication costs as well as the dielectric and conductor losses introduced by the decoupling structure even at frequencies in the decoupled regime create an inconvenience. We should note that a similar approach was reported in [19] in which a metamaterial spacer is employed to reduce the mutual coupling between two monopoles, each 27.8 mm in height, installed on a common ($2\,\lambda_0 \times 2\,\lambda_0$) aluminum ground plane and separated by 0.156 λ_0 (18 mm) at 2.6 GHz. The principle of operation is explained by considering the metamaterial acting as an artificial magnetic conductor (AMC). The decoupling structure consists of four existing design of a bidirectional AMC printed on a dielectric substrate with width-height-thickness dimensions of 64 × 33 × 10.5 mm.

Several defected ground planes [9–12], soft artificial electromagnetic surfaces, matching and decoupling networks [14–16], and neutralization lines [17] have received considerable recent research interest in both academia and industry. However, they were not applied specifically to vertical monopole antennas, the subject of this chapter. It should be noted that these traditional approaches suffer from:

a. Extremely narrow bandwidths since metamaterials, defected ground structures, and soft electromagnetic surfaces embrace inherently high-resonant structures;

b. Repeated periodic structures are employed periodically in 1-D, 2-D, or in 3-D using vias, which further complicates the design and increases the fabrication cost;

c. Insertion of many 2D cells (arranged in multiple rows and/or columns) and sometimes 3D unit cells [7, 8] between the antenna elements limits the empty space between the antenna elements and hence complicates the task of miniaturizing the MIMO antenna array; and

d. Distortion of the far-field radiation patterns and/or reduction the operating bandwidth [18–20] from these artificial structures.

To conclude, most of the related research utilized meta-surfaces inclusions inserted in the space *between* the antenna elements for which the mutual coupling is to be reduced. This is another major obstacle for applications involving extremely miniaturized antenna arrays, such as MIMO systems on handheld wireless devices. In this chapter, however, we report a novel approach in which the mutual coupling is reduced by inserting two vertical monopole antenna elements *inside* a rectangular conducting ring loaded with two conducting strips. This technique allows much smaller separations ($\lambda_o/16$), where λ_o is the free-space wavelength. Unlike the traditional techniques that are based on inserting artificial resonant structures between the radiating elements, the methodology reported in this chapter is based on *enclosing* the radiating elements by simple and versatile planar conducting structures, the performance of which exceeds the performance of traditional/existing design approaches.

2. MIMO array structure

We will elaborate on one embodiment involving a miniaturized two-element antenna array system composed of two vertical monopole antennas separated by 8 mm, corresponding to $\lambda_o/16$, where λ_o is the wavelength in free space at 2.4 GHz. The unique design scheme of the coupled array produces orthogonal radiation patterns even without the inclusion between the monopoles. Additionally, the decoupling structure is designed to reduce the mutual coupling and enhance isolation between the signals at the array terminals. A very low envelope correlation coefficient between the received signals is realized to accommodate operating with the different MIMO modes. To summarize, the design of the antenna array disclosed in this chapter seeks to achieve three functions: decoupling, descattering, and orthogonalization of the far-field radiation patterns in the boresight directions. Descattering in this context implies that the radiation pattern of each antenna element in the array is the same as the radiation pattern when in isolation.

To obtain a boresight oriented by 45° relative to that of a conventional vertical monopole, the design uses a partial reference ground plane instead of a traditional full ground as shown in **Figure 1**. The monopole is mounted on a dielectric substrate and is fed by a coplanar waveguide (CPW) circuit. The ground plane of the coplanar feeding circuit serves as a partial ground for the antenna as well. The height of the monopole h and the feeding circuit geometry are tuned such that the antenna operates at center frequency of 2.4 GHz serving applications in the ISM band. The coplanar waveguide feeding circuit is selected to have a characteristic impedance of 50 ohm at its input. The geometry, reflection coefficient, and radiation pattern of the single element are shown in **Figures 1–3**, respectively.

This behavior is exploited to construct an array consisting of two vertical monopoles, the radiation patterns of which are identical in shape but orthogonally oriented in space. The antenna array is formed by placing monopole 2 and its feeding circuit as a mirror image of monopole 1 around the y-z plane as depicted in **Figure 4**. Both antennas are assembled on a common substrate of a relative permittivity, ε_r = 10.2, and height, h_s = 1.27 mm.

When the two antennas are positioned in a close proximity, $\lambda_o/16$, in this design, the far-field radiation patterns will be distorted while the correlation coefficient

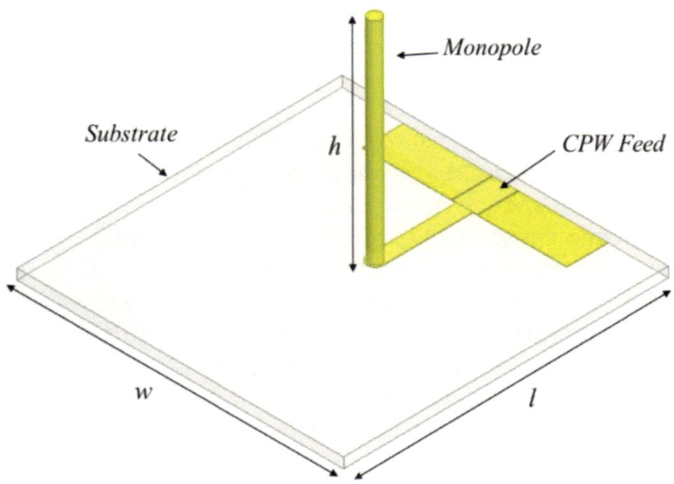

Figure 1.
Geometry of a single monopole antenna fed by CPW.

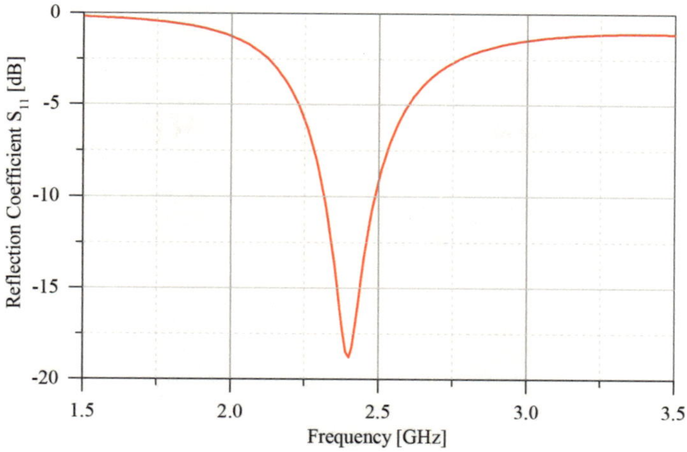

Figure 2.
Reflection coefficient of the monopole antenna fed by a CPW.

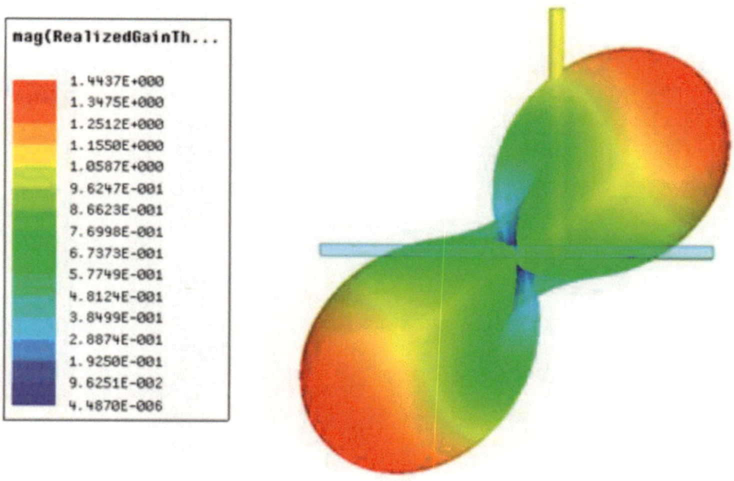

Figure 3.
Realized gain of the monopole antenna fed by a CPW.

increases due to mutual coupling. To investigate the mutual coupling effects, the array is simulated using the commercial software package, high-frequency structure simulator (HFSS) of ANSYS. The geometry of the design is shown in **Figure 5** and the dimensions are presented in **Table 1**.

Despite the reduction in transmission level between the feeding ports, the distribution of surface currents depicted in **Figure 6**, resulted from exciting port 1 while terminating port 2 with 50 ohms, shows the existence of significant interfering currents distributed on the CPW feeding of monopole 2. The current induced on monopole 2 and its feeding circuitry is due to the mutual coupling between the two antennas.

In **Figure 7**, we show the reflection coefficient, S_{11} and transmission coefficient, S_{21} plotted against frequency. The coupled array provides a bandwidth of 230 MHz over which the reflection coefficient is below −10 dB. At 2.4 GHz, the partial ground plane reduced S_{21} to −10 dB compared to −4 dB if the antennas share a common full ground plane and are separated by 8 mm. The radiation patterns in terms of the realized gain are plotted in **Figure 8** for two cases; in the first case, an

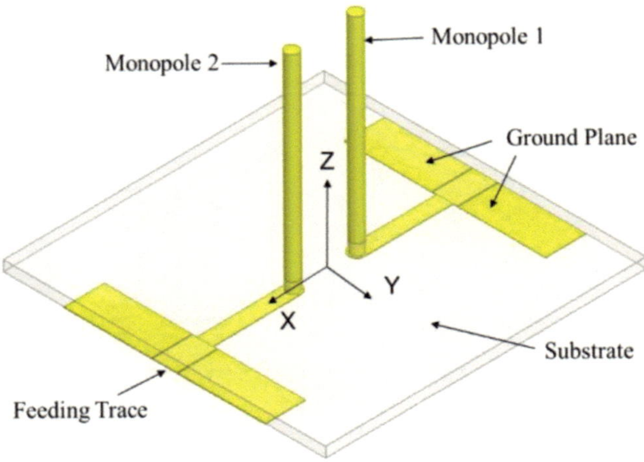

Figure 4.
The coupled monopole antenna array.

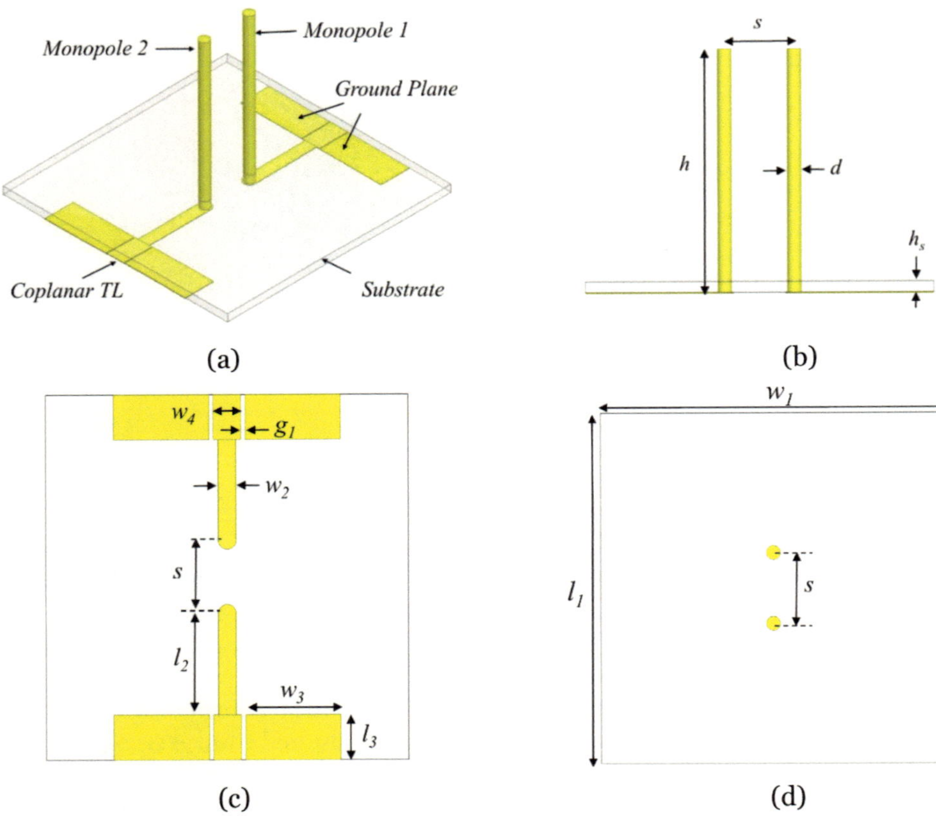

Figure 5.
Geometry of the coupled antenna array. (a) Prospective view, (b) Side view, (c) Bottom view and, (d) Top view.

excitation signal is injected into port 1, while the second port is terminated with its internal impedance, 50 ohms. In the second case, the excitation signal is injected into port 2, and port 1 is terminated by 50 ohms. Having the antennas placed extremely close to each other (8 mm at 2.4 GHz) resulted in disturbing the near

Monopoles separation (s)	8
Monopole height (h)	25
Monopole diameter (d)	1.6
Feeding probe length (l_2)	11
Feeding probe width (w_2)	2
Ground width (w_3)	10.85
Ground length (l_3)	5
Center strip width (w_4)	3
CPW gap width (g_1)	0.5
Substrate ($l_1 \times w_1 \times h_s$)	(40 × 40 × 1.27)
Dielectric constant (ε_r) of the substrate	10.2

Table 1.
Dimensions of the coupled antenna Array.

field and increasing the coupling level. This proximity causes the radiation patterns to be directed into the lower part below the plane containing the substrate.

The proposed MIMO antenna array consists of two monopole antennas positioned on a printed circuit board. The antennas are surrounded by a decoupling circuit printed on the top side of the substrate and a feeding circuit on the bottom side. For short, the design will be called the decoupled coplanar waveguide fed monopole antenna array (CPW-MAA). The substrate has a length and width of 40 × 40 mm, and a thickness of 1.27 mm.

In order to further enhance the performance of pattern diversity for the two-element antenna array and minimize the contribution of mutual coupling among the elements of the antenna array to the channel correlation, the mutual coupling effect needs to be reduced with minimal distortion to the radiation patterns of the individual elements, preferably such that the 3-D far-field patterns of the individual elements being identical in shape and orthogonal in space. This applies to the signals at the output of the antennas when transmitting and at the input of the feeding circuit when operating in the receive mode.

The decoupling structure, shown in **Figure 9**, consists of a simple 2D planar rectangular metallic ring and two tuning metallic strips printed on the surface of the substrate.

First, the performance of the CPW-MAA is compared against the coupled monopole antenna array shown in **Figure 5** using the same substrate. Furthermore, as a design alternative and to further validate the disclosed design concept, another substrate has been used with a dielectric constant ε_r of 6.15 and height, 1.27 mm. Each monopole is fed by a copper trace printed on the bottom side of the substrate. Each monopole makes a contact with the copper trace by extending the monopole body inside the substrate to reach the feeding copper trace at the bottom layer. A CPW at the bottom side of the substrate is used to deliver the signal to the antennas in case of transmitting and from the antennas in case of receiving.

The CPW consists of a central strip having a length of l_3 and a width of w_4 with the finite ground planes positioned on both sides of the central strip with a width of w_3. An SMA is utilized to launch the wave into the array. The impedance of the SMA is 50 ohm and is used to connect the antenna to a vector network analyzer used in the measurements. The array system dimensions are tuned to operate at a

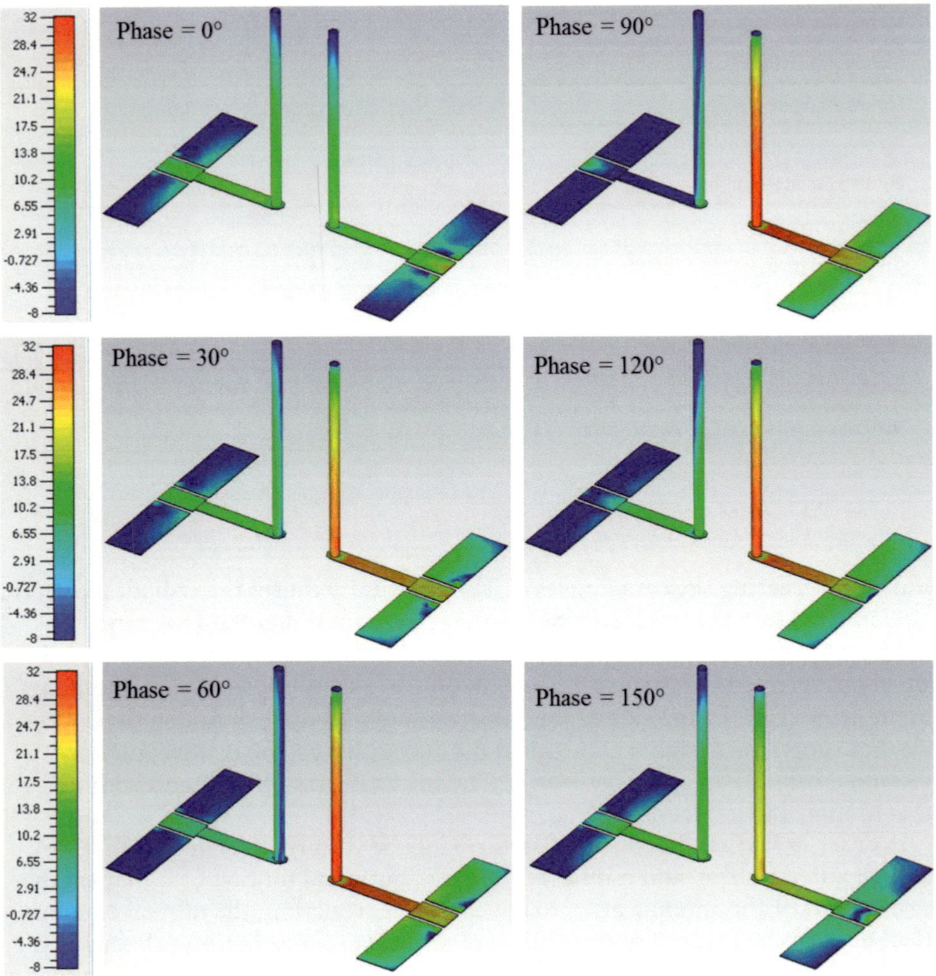

Figure 6.
Magnitude of surface current distribution for the coupled monopole antenna array, dB(1 A/m).

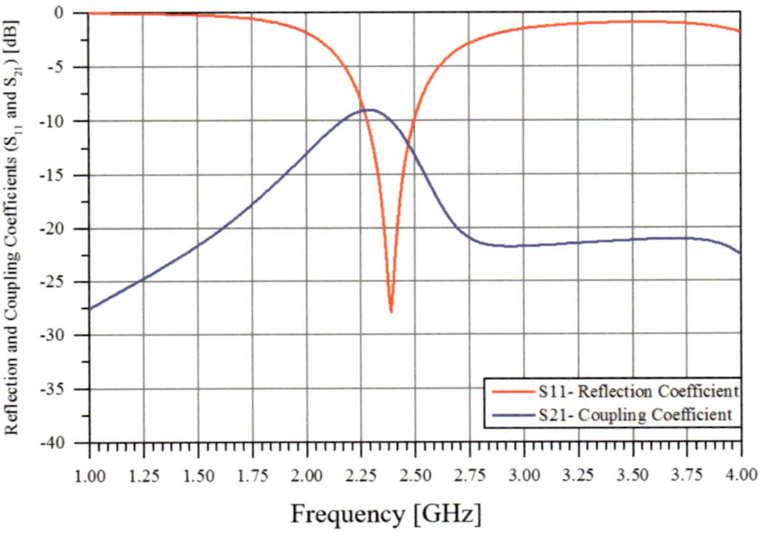

Figure 7.
Frequency response of the coupled monopole antenna.

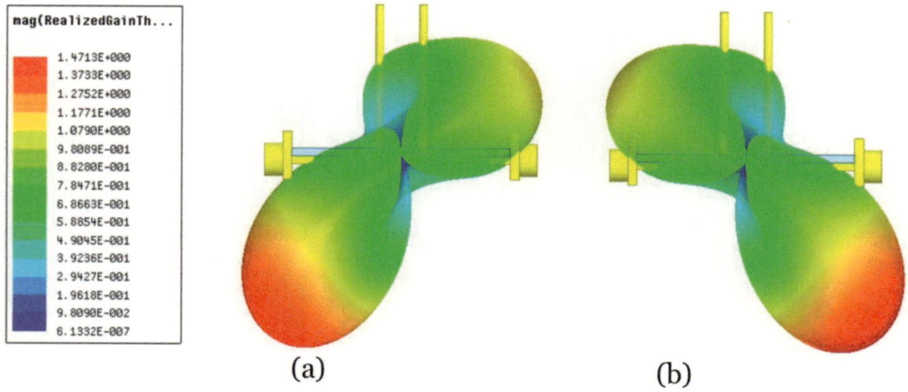

Figure 8.
Mutual coupling effects on the far-field radiation patterns. (a) Exciting port 1 and, (b) Exciting port 2.

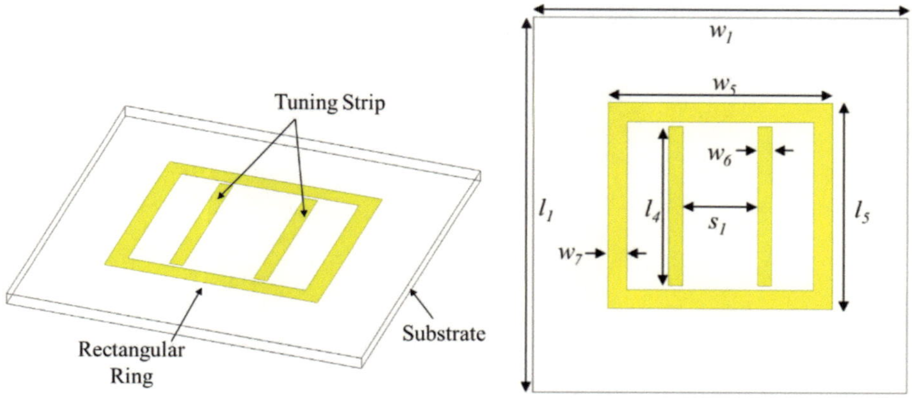

Figure 9.
Decoupling circuit.

2.4 GHz with a bandwidth of 260 MHz over which the reflection coefficients, S_{11} and $S_{22,}$ are less than −10 dB and transmission coefficients, S_{12} and S_{21} less than −20 dB. The two-element MIMO antenna array is shown in **Figure 10**, while dimensions are depicted in **Table 2** for the two substrates used where all dimensions are in millimeters.

Prototypes are fabricated and measured at the University of Arkansas at Little Rock's Antennas and Wireless Systems Research Laboratory (AWSRL). The prototype which is shown in **Figure 11** used Roger RO3210 substrate. The frequency response has been measured using E5071B and N5242A VNAs. The radiation patterns are measured inside an anechoic chamber. Using Roger RO3210 (dielectric constant ε_r of 10.2) as a substrate, and in order to demonstrate that adding this decoupling circuit isolates the input and output ports, the magnitude of the surface current dB (1 A/m) has been plotted for different excitation phase delays as depicted in **Figure 12**. A signal of 1 W is assigned through port 1 to excite the first monopole, the monopole in the right side of the array, while port 2 is terminated by its internal impedance of 50 ohm. It seems quite evident that the amount of current coupled to the feeding circuit of monopole 2 at different phase delays is very low, minimum at −8 dB (1 A/m) (blue color), compared to other excited monopole, where a high current is observed with a maximum value at 32 dB (1 A/m) (red color). This contrast of 40 dB (1 A/m) reveals the high isolation level achieves due

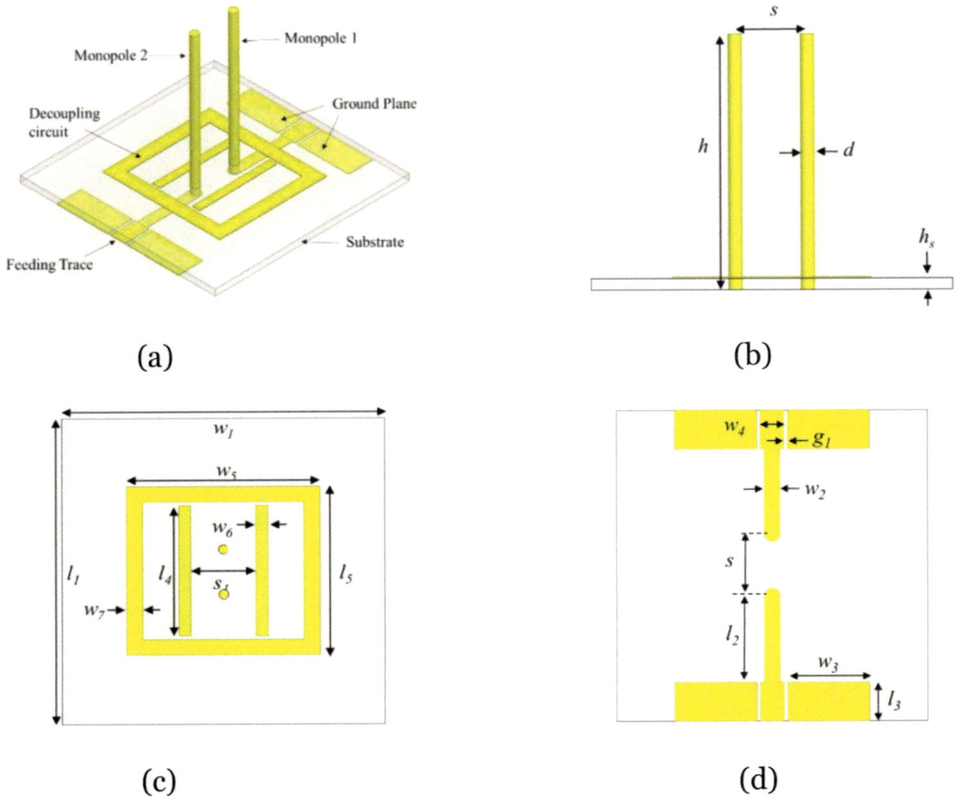

Figure 10.
The decoupled CPW-MAA. (a) Prospective view, (b) Side view, (c) Top view, (d) Bottom view.

Dimension defined name	Substrate RO3210 ($\varepsilon_r = 10.2$)	Substrate RO3006 ($\varepsilon_r = 6.15$)
Monopoles separation (s)	8	8
Monopole height (h)	28	27
Monopole diameter (d)	1.6	1.6
Feeding probe length (l_2)	11	11
Feeding probe width (w_2)	2	2
Ground width (w_3)	10.5	10.8
Ground length (l_3)	5.0	5.0
CPW trace width (w_4)	3.0	3.0
CPW gap width (g_1)	0.5	0.2
Rectangular ring outer length (l_5)	22	22
Rectangular ring outer width (w_5)	24	30
Rectangular ring trace width (w_7)	2.5	2.5
Tuning strip length (l_4)	16	16
Tuning strip width (w_6)	1.5	4
Tuning strips separation (s_1)	6.5	8
Substrate ($l_1 \times w_1 \times h_s$)	(40 × 40 × 1.27)	(40 × 40 × 1.27)

Table 2.
Dimensions of the CPW-MAA for two substrates.

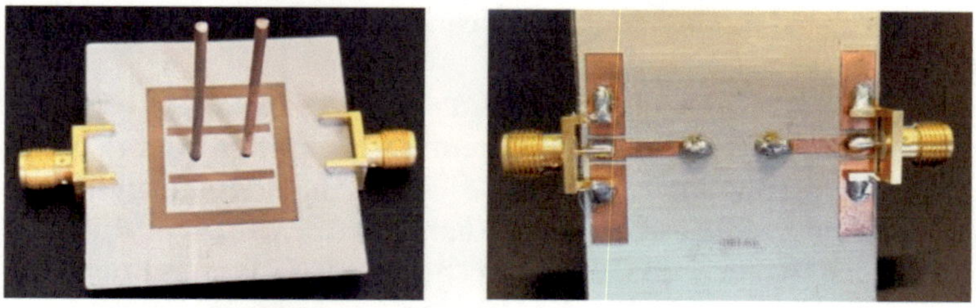

Figure 11.
The decoupled CPW-MAA and decoupled MS-MAA prototypes.

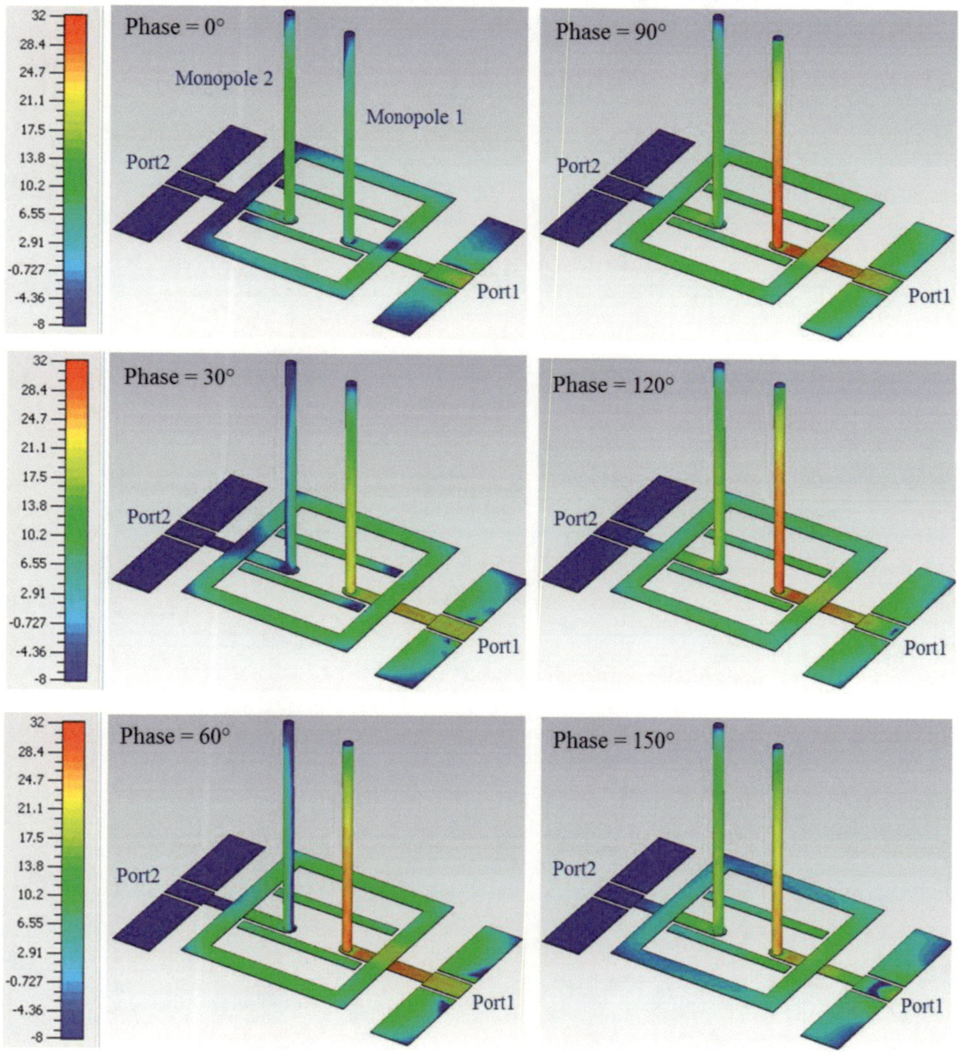

Figure 12.
Two monopole antenna array with decoupling network; surface current distribution (magnitude) dB(1 A/m).

to the decoupling network compared to the coupled current before adding the decoupling network shown in **Figure 6**.

The simulated and measured results of the scattering parameter, S_{11} are presented in **Figure 13**. The amount of coupling expressed in terms of scattering parameter, S_{21} is depicted in **Figure 14**. These figures reveal good agreement between the simulated models and the measurements using the fabricated prototype.

Introducing the decoupling network resulted in a reduction of −19 dB in mutual coupling level, from −4 to −23 dB. The antenna array system offers a bandwidth of 10%, in which the coupling coefficient, S_{21} is less than −20 dB and 14% over which the reflection coefficient, S_{11} is less than −10 dB. The slight shift of 2% in the frequency at which the measured coupling coefficient S_{21} is at its minimum in the bandwidth of interest compared to simulated results is due to the variations associated with the fabrication of the prototype and the limited precision of simulations and measurements. The fabricated prototype also provides a wider bandwidth. This is due to the dielectric and metallic losses inherent in the substrate and dipoles, which lead to reduced *Q* factor.

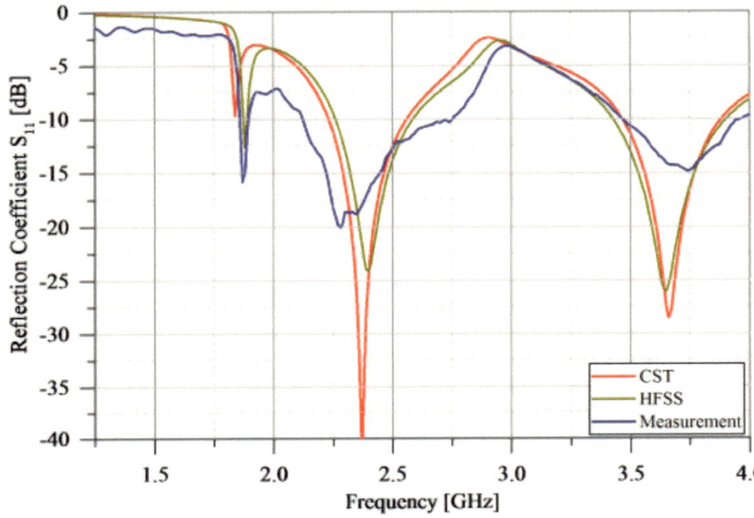

Figure 13.
Reflection coefficient S_{11}, simulation versus measurements.

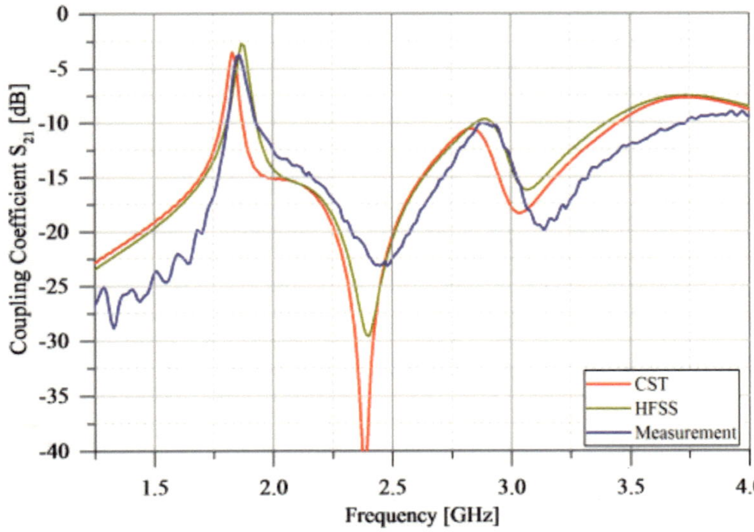

Figure 14.
Coupling coefficients S_{21}, simulation versus measurements.

Table 3 summarizes the performance metrics of the simulated models and the measured prototype. In this table, f_L and f_H denote the lower and upper frequency limits of the operating bandwidth, respectively. The array meets and exceeds the bandwidth requirement of 802.11 Wi-Fi systems operating in the ISM band at 2.4 GHz.

The maximum value of the total realized gain is plotted versus frequency in **Figure 15**. Another advantage of the proposed antenna array is the enhancement achieved in the realized gain and hence the efficiency. The total realized gain of a single element varies over the range from 1.6 to 1.69 dB over the frequency range 2.25–2.52 GHz over which S_{11} and S_{21} are less than −10 and −20 dB, respectively. This gain is obtained when one port is excited while the other is terminated with its characteristic impedance. An improvement of 2.1 dB is observed in the realized gain with the proposed antenna array as compared to a conventional array antenna consisting of two monopoles separated by $\lambda_0/16$ over a $\lambda_0 \times \lambda_0$ full ground plane. It is

Performance metric	Conventional array	HFSS	CST	Measured
S_{11} (min) dB	−8	−24	−40	−20
S_{22} (min) dB	−8	−23	−40	−33
S_{21} (min) dB	−4	−29	−45	−23
f_L ($S_{11}, S_{22} < -10$ dB) GHz	—	2.25	2.22	2.12
f_H ($S_{11}, S_{22} < -10$ dB) GHz	—	2.58	2.56	2.74
f_L ($S_{12}, S_{21} < -20$ dB) GHz	—	2.28	2.26	2.32
f_H ($S_{12}, S_{21} < -20$ dB) GHz	—	2.51	2.51	2.58
BW ($S_{11}, S_{22} < -10$ dB) GHz	—	0.33	0.34	0.62
BW ($S_{21}, S_{12} < -20$ dB) GHz	—	0.23	0.25	0.26
% BW ($S_{11}, S_{22} < -10$ dB)	—	13.6%	14.2%	25.5%
% BW ($S_{21}, S_{12} < -20$ dB)	—	9.6%	10.4%	10.6%

Table 3.
Decoupled array performance at 2.4 GHz in terms of scattering parameters (dB).

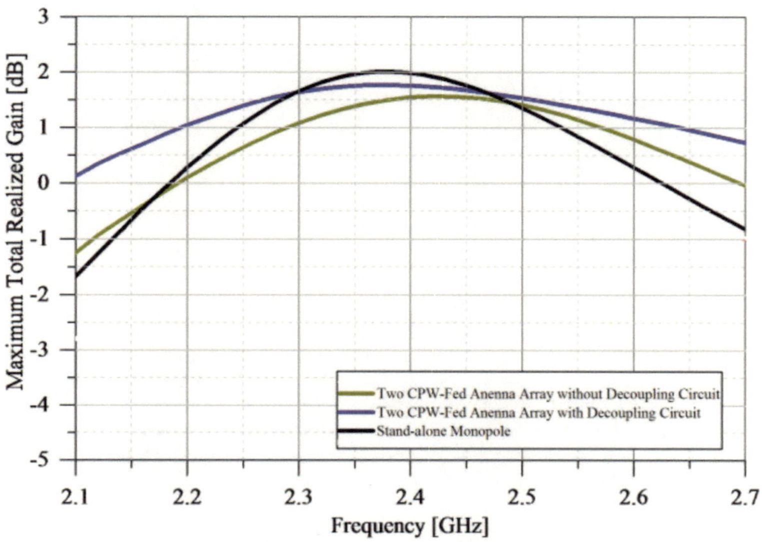

Figure 15.
Maximum realized gain.

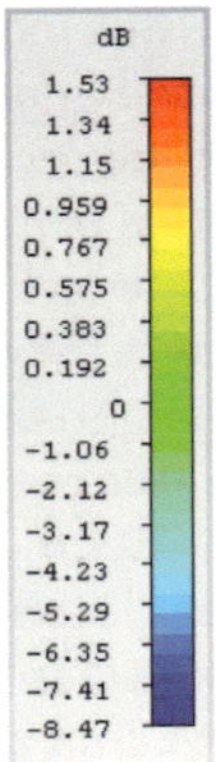

(a) Realized Gain Scale (dB)

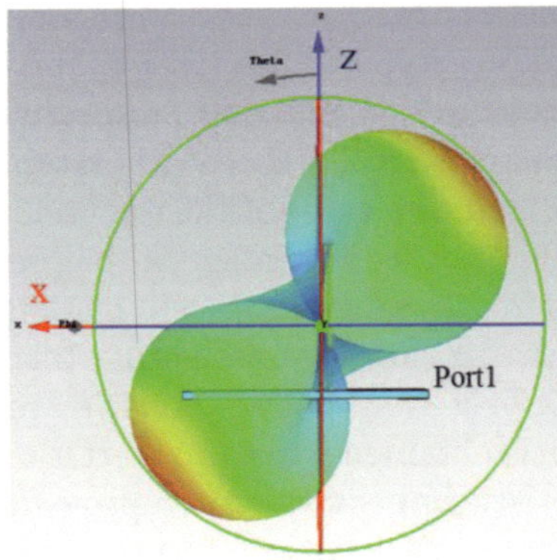

(b) Model 1 (Stand-alone Monopole)

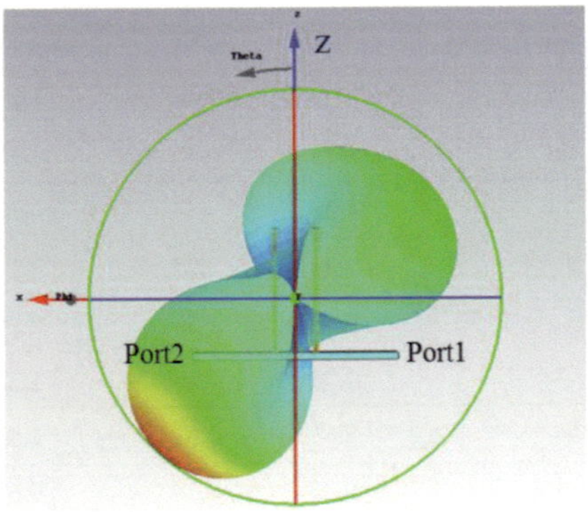

(c) Two monopoles antenna array: Exciting port 1

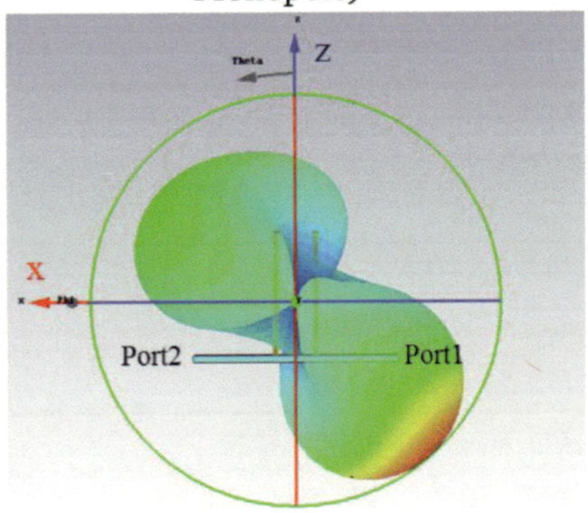

(d) Two monopoles antenna array: Exciting port 2

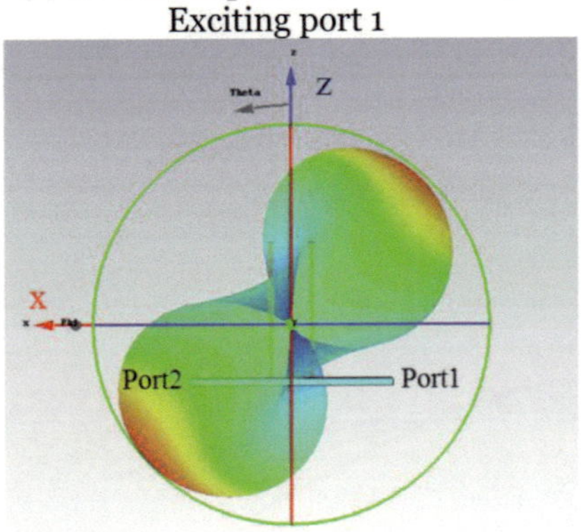

(e) Decoupled CPW-MAA: Exciting port 1

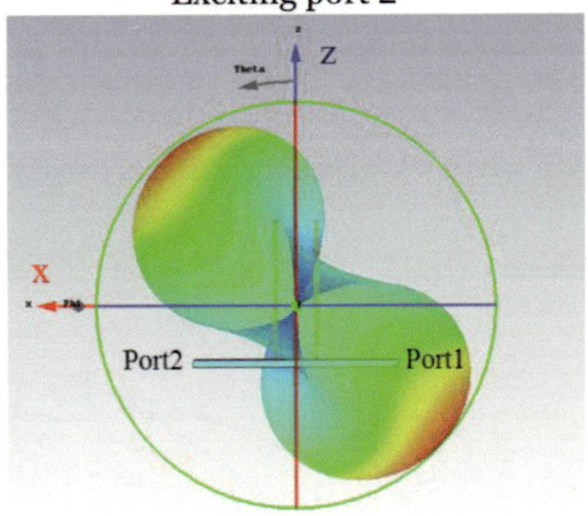

(f) Decoupled CPW-MAA: Exciting port 2

Figure 16.
(a) The realized gain scale. (b) Realized gain for a stand-alone monopole, while (c) and (d) are for the coupled monopole antenna array, then (e) and (f) are for the decoupled CPW-MAA. Examining the upper right section of the 3-D pattern of model 2, ($\pi \le \phi \le 2\pi$, and $\theta \le \pi/2$) shows red spot, which indicates high gain which is reduced for the decoupled CPW-MAA, where more power is directed in the lower left side ($0 \le \phi \le \pi$ and $\theta \ge \pi/2$). This distortion has been corrected in the disclosed antenna array due to the low mutual coupling caused by adding the decoupling network. Furthermore, a descattered pattern is realized. **Figure 17** *shows a comparison between simulated and measured co/cross radiation patterns of the decoupled CPW-MAA (***Figure 18***).*

noteworthy that the maximum value of the total realized gain of the array system is close to that of a stand-alone monopole antenna.

The 3-D radiation patterns in terms of the realized gain are shown in **Figure 16**.

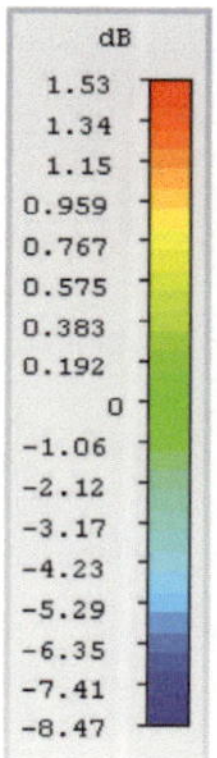

(a) Realized Gain Scale (dB)

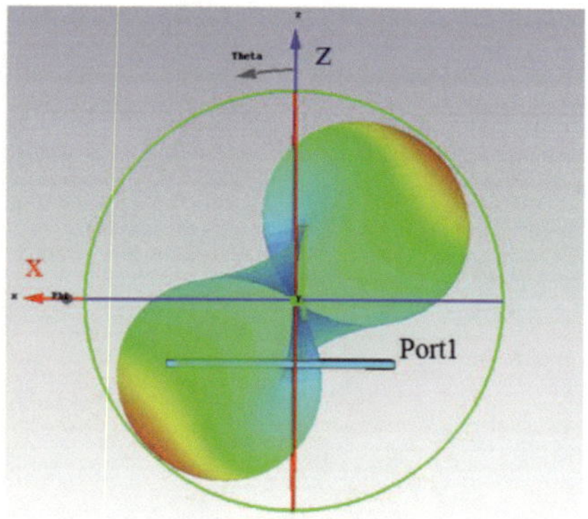

(b) Model 1 (Stand-alone Monopole)

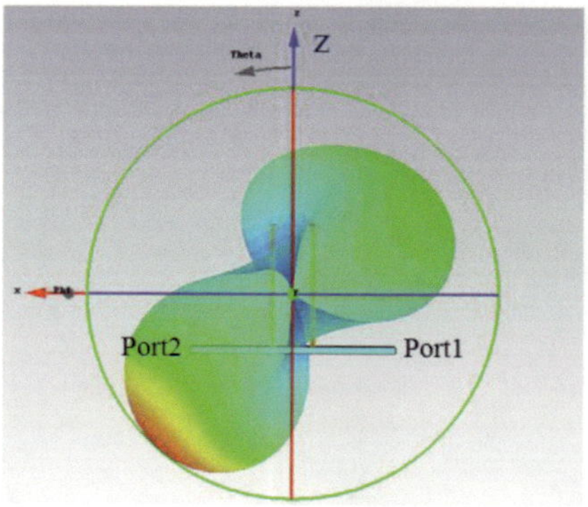

(c) Two monopoles antenna array: Exciting port 1

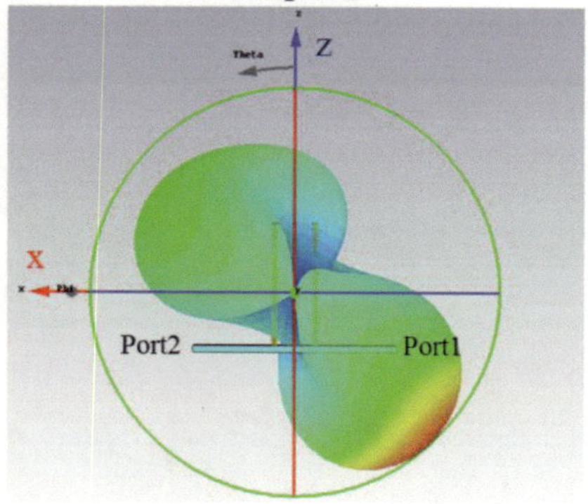

(d) Two monopoles antenna array: Exciting port 2

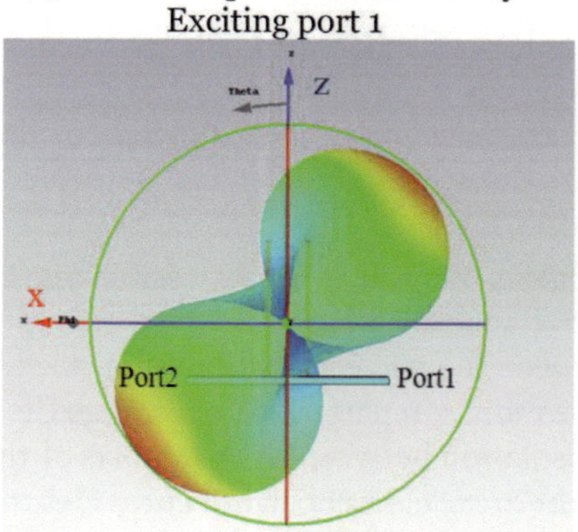

(e) Decoupled CPW-MAA: Exciting port 1

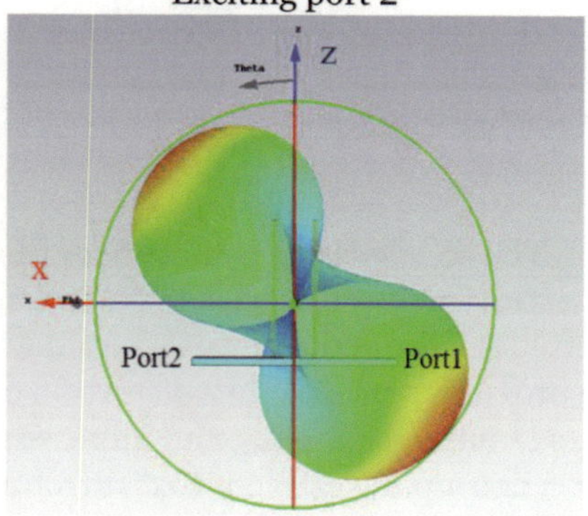

(f) Decoupled CPW-MAA: Exciting port 2

Figure 17.
Realized gain; scale (a), stand-alone monopole (b), two-monopole array without decoupling structure (c) and (d), decoupled CPW-MAA (e) and (f).

3. Conclusions

A compact MIMO antenna array with identical yet orthogonal radiation patterns and high isolation level is reported in this chapter. The shape and size of the partial ground plane and its distance from the monopole allowed the boresight of the radiation pattern of the isolated monopole to be oriented at an elevation angle

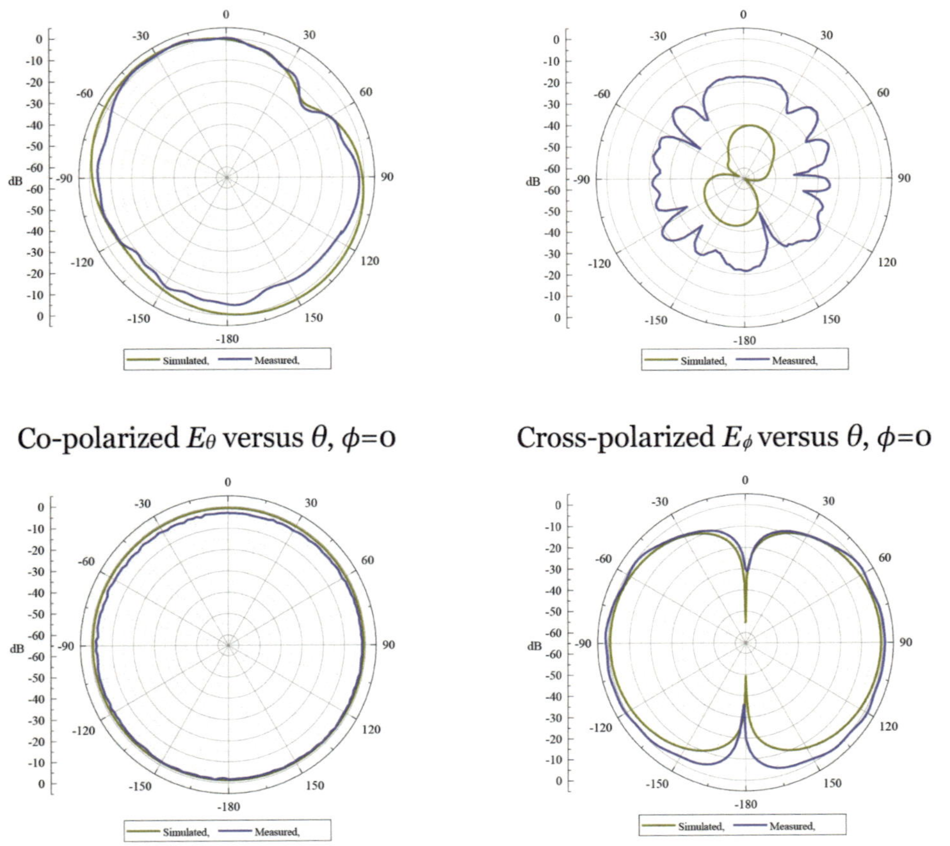

Co-polarized E_θ versus θ, ϕ=0

Cross-polarized E_ϕ versus θ, ϕ=0

Co-polarized E_θ versus ϕ, $\theta = \pi/2$

Cross-polarized E_ϕ versus ϕ , $\theta = \pi/2$

Figure 18.
Realized gain; simulation versus measurements.

of 45°. The 45° elevation achieved for the single monopole made it possible to place two vertical monopoles back to back to achieve an array with orthogonal radiation patterns. However, the mutual coupling between the two elements when placed in close proximity resulted in a distortion of the radiation patterns. To mitigate the effects of mutual coupling and increase the isolation between the two ports of the array, a decoupling element that surrounds the array elements which provides more flexibility on how close the elements need to be from each other. The dimensions of the rectangular ring and the tuning strips have been tuned to achieve the desired isolation level and restore the radiation pattern of the isolated element. The CPW-MAA with orthogonal far-field radiation patterns opens new opportunities for modern handheld devices and/or fixed wireless access points implementing MIMO techniques. Unlike existing design approaches that are based on inserting artificial resonant structures between the radiating elements, the design presented in this chapter encloses the radiating elements by simple and versatile planar conducting structures. The proposed design is demonstrated by simulations and measurements to significantly reduce the mutual coupling when the monopoles are spaced by 8 mm (λ_0/15.3). The performance achieved in terms of impedance matching, isolation, gain, bandwidth, and radiation pattern is comparable to network decoupling and parasitic approaches and better in terms of bandwidth, size, and complexity in comparison to existing designs which employ periodic AMC. A matching bandwidth of 0.634 GHz over which the reflection coefficient, S_{11} is less than −10 dB, and a 0.246 GHz bandwidth over which the transmission coefficient, S_{21} is less than −20 dB have been achieved. A total realized gain of 1.69, 1.95, and 1.68 dB are achieved at 2.26, 2.4, and 2.52 GHz, respectively.

Author details

Hussain Al-Rizzo*, Ayman A. Isaac, Sulaiman Z. Tariq and Samer Yahya
Department of Systems Engineering, George W. Donaghey College of Engineering and Information Technology, University of Arkansas at Little Rock, Little Rock, AR, USA

*Address all correspondence to: hmalrizzo@ualr.edu

References

[1] Clerckx B, Oestges C. MIMO Wireless Networks Channels, Techniques and Standards for Multi-Antenna, Multi-User and Multi-Cell Systems. 2nd ed. Amsterdam: Academic Press; 2017

[2] MMda S, Monteiro FA. MIMO Processing for 4G and beyond: Fundamentals and Evolution. Boca Raton: CRC Press; 2016

[3] Hampton JR. Introduction to MIMO Communications. Cambridge: Cambridge University Press; 2014

[4] Sievenpiper D, Zhang L, Broas R, Alexopolous N, Yablonovitch E. High-impedance electromagnetic surfaces with a forbidden frequency band. IEEE Transactions on Microwave Theory and Techniques. 1999;**47**(11):2059-2074

[5] Yablonovitch E. Circuit and Method for Eliminating Surface Currents on Metals. 2001

[6] Kawaguchi K, Sugimoto Y, Kondo A, Yukumatsu M. Antenna Apparatus Having Patch Antenna. 2016

[7] Dandekar K, Mookiah P. MIMO Antenna Arrays Built on Metamaterial Substrates. 2016

[8] Mookiah P, Dandekar K. Bi-Directional Magnetic Permeability Enhanced Metamaterial (MPEM) Substrate for Antenna Miniaturization. 2015

[9] Murch R, Chiu C-Y, Rowell C. Systems and Methods Using Ground Plane Filters for Device Isolation. 2009

[10] Wu T-L, Hsu Y-H, Tsai C-H. Defected Ground Structure with Shielding Effect. 2010

[11] Abushamleh S, Al-Rizzo H, Kishk A, Abbosh A, Khaleel H. Miniaturized thin soft surface structure using metallic strips with ledge edges. Progress In Electromagnetics Research. 2014;**57**:221-232

[12] Abushamleh S, Al-Rizzo H, Abbosh A, Kishk A. Mutual coupling reduction between two patch antennas using a new miniaturized soft surface structure. In: 2013 IEEE International Symposium on Antennas and Propagation and USNC-URSI National Radio Science Meeting. 2013

[13] Kim C, Kim J, Choi C-G, Beak G, Park Y, Ahn H, et al. MIMO Antenna Having Parasitic Elements. 2013

[14] Zhao L, Wu K-L. A dual-band coupled resonator decoupling network for two coupled antennas. IEEE Transactions on Antennas and Propagation. 2015;**63**(7):2843-2850

[15] Zhao L, Yeung LK, Wu K-L. A coupled resonator decoupling network for two-element compact antenna arrays in Mobile terminals. IEEE Transactions on Antennas and Propagation. 2014;**62**(5):2767-2776

[16] Zhao L, Qian K-W, Wu K-L. A cascaded coupled resonator decoupling network for mitigating interference between two radios in adjacent frequency bands. IEEE Transactions on Microwave Theory and Techniques. 2014;**62**(11):2680-2688

[17] Håkansson M, Ying Z. Multiple-Input Multiple-Output (MIMO) Multi-Band Antennas with a Conductive Neutralization Line for Signal Decoupling. 2014

[18] Bait-Suwailam MM, Boybay MS, Ramahi OM. Electromagnetic coupling reduction in high-profile monopole antennas using single-negative magnetic Metamaterials for MIMO applications. IEEE Transactions on Antennas and Propagation. 2010;**58**(9):2894-2902

[19] Ferrer PJ, González-Arbesú JM, Romeu J. Decorrelation of two closely spaced antennas with a metamaterial AMC surface. Microwave and Optical Technology Letters. 2008;**50**(5):1414-1417

[20] Dadashzadeh G, Dadgarpour A, Jolani F, Virdee B. Mutual coupling suppression in closely spaced antennas. IET Microwaves, Antennas and Propagation. 2011;**5**(1):113

[21] Capolino F. Metamaterials Handbook. Boca Raton: CRC Press; 2009

[22] Lau BK, Andersen JB. Simple and efficient decoupling of compact arrays with parasitic Scatterers. IEEE Transactions on Antennas and Propagation. 2012;**60**(2):464-472

Chapter 2

Low-Profile Metamaterial-Based Adaptative Beamforming Techniques

Chung-Tse Michael Wu and Pai-Yen Chen

Abstract

In this chapter, we will review recent research advances on beamforming and spatial multiplexing techniques using reconfigurable metamaterials (MTMs) and metasurfaces. This chapter starts by discussing basic principles and practical applications of transmission line-based metamaterials and planar metasurfaces, followed by their active versions that enable novel smart antennas with beam steering and beamshaping functions. We include detailed descriptions of their practical realizations and the integration with circuits and the radio-frequency (RF) frontend, which are used to adaptively and dynamically manipulate electromagnetic radiation. We summarize the state-of-the-art MTM/metasurface-based beamforming techniques and provide a critical comparison for their uses in the RF-to-millimeter-wave range in terms of cost, reconfigurability, system integratability and radiation properties. These techniques are expected to pave the way for the massive deployment of communication, radar, remote sensing and medical and security imaging systems.

Keywords: beamforming, composite right/left-handed transmission lines, leaky wave antennas, metamaterials, metasurfaces

1. Introduction

Metamaterials (MTMs) are artificial composite structures having anomalous electromagnetic properties that are not found in naturally occurring materials. Examples include the negative refractive index [1–3] and the terahertz (THz) and optical magnetism [4]. The word “meta” is sourced from a Greek word that means beyond or after. Metamaterials are generally constituted by arrays of subwavelength inclusions with induced electric and/or magnetic dipole moment, such as magnetodielectric spheres. The periodicity between neighboring inclusions is much smaller than the wavelength of impinging light and dominant Bloch modes. The first 3D MTM was realized in the microwave regime using arrays of metal wires and split-ring resonators (SRRs) [2–4]. For the first time, this structure has been demonstrated to exhibit a simultaneously negative permittivity and permeability, resulting in a negative index of refraction. A negative-index or double-negative (DNG) MTM is generally named as a “left-handed” medium, which supports the electromagnetic wave propagation with an exotic negative phase velocity and negative refraction. In a left-handed medium, the Snell’s law is inverted and evanescent waves could be amplified to make a "perfect lens" that enables sub-diffraction imaging [2, 3].

The left-handed propagation properties can also be realized using the printed-circuit transmission line (TL), which eases the fabrication complexity of metamaterials at high frequencies [5–7]. The TL-based MTMs, also termed as composite right/left-handed transmission lines (CRLH-TLs), have gained popularity in RF and microwave bands, owning to their low cost, compactness and compatibility with the printed circuit technology. Moreover, the TL structures based on non-resonant lumped elements can provide broader bandwidth and lower power dissipations compared with those bulky metamaterials composed of 3D arrays of resonant inclusions. The planar TL MTMs also allow the integration with active and/or passive tuning components, such as varactors or field-effect transistors (FETs), for making adaptive and spectral/spatial-reconfigurable electromagnetic responses. The TL MTMs therefore provide a useful platform for observing the phenomena of negative material properties and for reaching various metamaterial-enabled applications. The exotic guided- and leaky-wave properties in the CRLH-TL medium, associated with the negative phase velocity, have been proposed to realize different kinds of planar microwave circuit blocks, such as the multi-band and enhanced bandwidth power combiners/splitters, resonators, couplers [8], phase shifters [9], bandpass filters [10] and subwavelength focusing devices [11, 12]. Furthermore, the fast wave and dual RH-LH characteristics existing in the periodic CRLH-TL structures have been exploited to excite leaky wave radiations and to implement one-dimensional (1D) and two-dimensional (2D) leaky-wave antennas (LWAs) with superior radiation properties and tunability [13–16]. These metamaterial-based leaky-wave antennas can not only be compact and low-profile, but also achieve a directive beam that can be steered from backfire, broadside to endfire. Such a wide-angle beam scanning is not available in conventional uniform or periodic leaky-wave antennas, which require complex and narrow-band feeding networks [17]. Most interestingly, by loading an MTM LWA with tuning elements, such as p-i-n or varactor switching diodes [18], it is possible to achieve fast and frequency-locked beamscanning, which is by the external biasing circuitry. Consequently, the beam scanning can function properly at a specific operating frequency, which is of great interest for the spatial channelizing in the modern communication systems.

RF and microwave beamforming techniques have drawn intensive research interest in 5G wireless communication [i.e., spatial multiplexing multiple access (SDMA) and space shift keying (SSK)], microwave imaging and radar for the directional and reconfigurable RF signal transmission or reception. Here, we note that the beamforming technique can be used at both the transmitting and receiving ends to achieve spatial selectivity. Conventional beamforming systems are generally based on mechanically steered antennas, with the assistance of motors and gimbals, which, however have several disadvantages, such as relatively large size and weight, as well as a slow beam scanning rate. Alternatively, phased arrays consisting of multiple active antennas are also commonly used to launch electromagnetic waves to a particular direction, by means of properly phasing the wavefront of individual antennas to cause constructive or destructive interference. However, phased arrays are composed of lots of radiating elements and at least one stage of amplifier and phase shifter, and are expensive to purchase due to their technical complexity.

In this chapter, we will present an overview of TL-based MTMs and their applications in smart and reconfigurable antennas. We will review different kinds of recently developed planar MTM and metasurface antennas with beamforming and beamshaping functions, as well as their practical applications. We begin by discussing the background of TL MTMs that lead to the development of printed

LWAs, and then introduce other state-of-the-art surface antenna technologies, such as the holographic antennas based on metasurfaces.

2. Transmission line metamaterials

Figure 1a illustrates the unit cell of a conventional lossless right-handed (RH) transmission line (TL) model, where the equivalent circuit contains series inductance (L_R) and shunt capacitance (C_R). The unit cell is assumed to represent a small section of the transmission line, much less than one quarter of the guided wavelength. The equivalent MTM constitutive parameters, such as permittivity and permeability can be obtained by mapping the Maxwell's equation onto the telegrapher's Equation [19]. As such, the propagation constant (β) can be derived in terms of the series impedance (Z') and shunt admittance (Y'), in which $Z' = j\omega L_R$ and $Y' = j\omega C_R$ [5]:

$$j\beta = \sqrt{Z'Y'} = j\omega\sqrt{L_R C_R} \tag{1}$$

From (1), we obtain a dispersion relationship that is linearly proportional to the frequency, as illustrated in (2) and **Figure 1b**:

$$\beta_{RH} = \omega\sqrt{L_R C_R} \tag{2}$$

On the other hand, a left-handed (LH) TL can be obtained by interchanging the inductor and capacitor in the RH TL unit cell. As illustrated in **Figure 2a**, the series impedance now is changed to a capacitance, whereas the shunt admittance becomes an inductance. Using a similar derivation as shown in (1), by replacing Z' with $1/j\omega C_L$ and Y' with $1/j\omega L_L$, the propagation constant can be expressed as:

$$\beta_{LH} = -\frac{1}{\omega\sqrt{L_L C_L}} \tag{3}$$

The negative sign of the propagation constant indicates the phase velocity is opposite to the direction of that for a RH-TL, which is governed by the "left-hand" rule. This left-handed (LH) TL is regarded as a kind of planar MTM, as it has novel negative-phase-velocity property that can only be artificially engineered. The corresponding dispersion diagram of such left-handed transmission line is shown in **Figure 2b**.

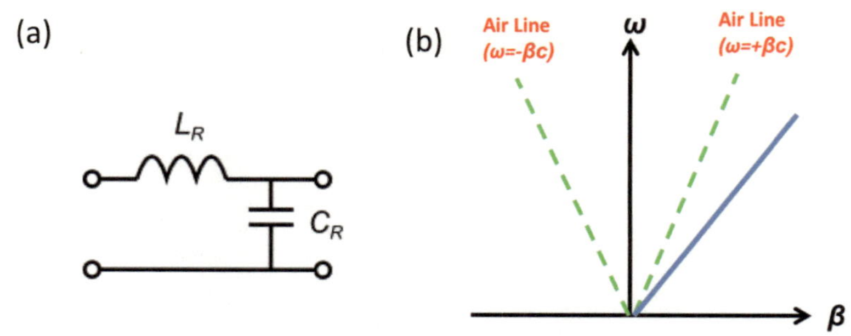

Figure 1.
(a) Unit cell and (b) dispersion diagram of a right-handed (RH)-TL model; the dashed lines represent the air lines.

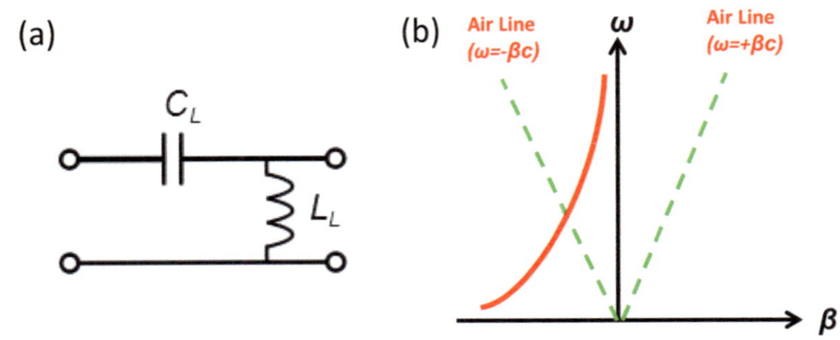

Figure 2.
(a) Unit cell and (b) dispersion diagram of a left-handed transmission line model; the dashed lines represent the air lines.

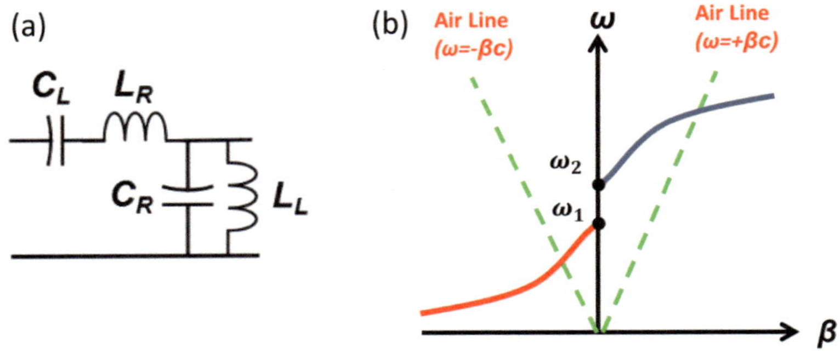

Figure 3.
(a) Unit cell and (b) dispersion diagram of a composite right/left-handed (CRLH)-TL model; the dashed lines represent the air lines.

Unlike the conventional case, the dispersion of a LH-TL is nonlinear and is always negative; the dashed lines represent the air lines. Nevertheless, in practice, due to the inevitable parasitic effect of host medium, it is rather difficult to realize the perfect left-handed TL MTMs. Indeed, the ordinary RH wave propagation is supported at higher frequencies. Therefore, a more generalized circuit model as shown in **Figure 3a** must be adopted to accommodate both RH and LH propagations. This configuration is the so-called composite right/left-handed transmission line (CRLH-TL). The dispersion relation can be derived using the same procedure by letting $Z' = j\omega L_R + 1/j\omega C_L$ and $Y' = j\omega C_R + 1/j\omega L_L$. After some mathematical manipulations, the propagation constant can be written as:

$$\beta_{CRLH} = s(\omega)\sqrt{\omega^2 L_R C_R + \frac{1}{\omega^2 L_L C_L} - \left[\frac{L_R}{L_L} + \frac{C_R}{C_L}\right]} \tag{4}$$

where

$$s(\omega) = \begin{cases} -1, \omega < \omega_2 = min\left\{\frac{1}{\sqrt{L_R C_L}}, \frac{1}{\sqrt{L_L C_R}}\right\} \\ 1, \omega > \omega_1 = max\left\{\frac{1}{\sqrt{L_R C_L}}, \frac{1}{\sqrt{L_L C_R}}\right\} \end{cases} \tag{5}$$

Figure 3b plots the corresponding dispersion curve, from which we can observe that the CRLH-TL can support both of the LH mode ($\beta < 0$) in the low frequency region and the RH mode ($\beta > 0$) in the high frequency region. Depending on the operating frequency, CRLH-TLs can have different LH or RH characteristics. In addition, if we define two air lines that have a slope of speed of light c, we can divide the dispersion diagram into two regions. The region sandwiched by the two air lines is called the *fast wave* region, as the phase velocity ($v_p = \omega/\beta$) is greater than c. On the other hand, the rest is defined as the *slow wave* region as v_p is slower than c. CRLH-TLs behave differently in these two regions. As will be shown in the next section, CRLH-TLs will radiate in the fast wave region, corresponding to a "leaky" mode, which is necessary for a LWA. On the other hand, when operated in the slow wave region, corresponding to a "guided" mode, CRLH-TLs behave like waveguiding structures with ignorable radiation loss. The dual RH/LH and guided-to-leaky wave characteristics make CRLH-TLs fascinating structures offering significantly more flexibility in the design of guided-wave and radiated-wave microwave components.

Figure 4 shows the comparison of dispersion characteristics between a CRLH-TL and a conventional RH-TL. As discussed in the previous section, the dispersion region can be divided into radiation (leaky wave) and guided wave regions. Moreover, depending on the polarity of the propagation constant β, the dispersion diagram can be categorized to right-handed ($\beta > 0$) and left-handed ($\beta < 0$) regions. A conventional RH-TL will lie in the RH guided region, as it has a positive β ($\beta > \omega/c$ or $v_p < c$) and will not radiate. On the other hand, the dispersion curve of CRLH-TL continues across all the four regions in the dispersion diagram, which are LH guided, LH radiated, RH guided and RH radiated wave regions. As a result, CRLH-TL structures provide flexibility to tailor the electromagnetics responses at will. The band gap at $\beta = 0$ ($v_p \to \infty$) will disappear, if the following condition is satisfied:

$$\sqrt{\frac{L_R}{C_R}} = \sqrt{\frac{L_L}{C_L}} \text{ or } L_R C_L = L_L C_R \tag{6}$$

We call this a "balanced" condition for a CRLH-TL. A balanced CRLH-TL does not have a band gap and therefore has a smooth transition at the center frequency where the propagation constant is zero. This property is especially useful when it operates as a leaky wave antenna, as it allows a continuous frequency-dependent beam-scanning capability.

If the CRLH-TL operates in the radiated or fast wave region with $\beta < \omega/c$ or $v_p > c$, it will leak out power to the free space as the wave travels along the transmission line. Therefore, a leaky wave antenna is sometimes referred as a fast-wave traveling wave antenna.

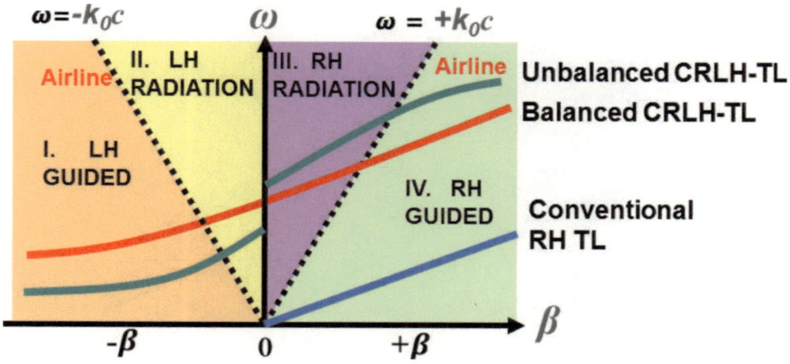

Figure 4.
Dispersion diagram of a composite right/left-handed transmission line model.

To further explain the leaky-wave characteristics, **Figure 5** illustrates the relationship between the vertical wavenumber k_z, the free space wavenumber k_0 and the modal propagation (phase) constant inside the wave guiding structure β_x. Here k_z can be expressed as:

$$k_z = \sqrt{k_0^2 - \beta_x^2} \tag{7}$$

When the wave propagates through the leaky wave structure with a complex wavenumber $\beta_x - j\alpha$, the fields outside the structure have the general form [17]:

$$\varphi(x,z) = \varphi_0\left(e^{-\alpha z}e^{-j\beta_x x}\right)e^{-jk_z z} \tag{8}$$

The attenuation constant $\alpha = \alpha_{loss} + \alpha_{rad}$, where α_{loss} is due to the material (dielectric and conduction) loss inside the guiding structure and α_{rad} is due to the radiative energy leakage (leakage rate). Therefore, one can observe that if k_z is a real number, that is, $k_0 > \beta_x$, the wave can propagate along the z-direction and hence the leakage radiation occurs. This is consistent with the aforementioned statement that when the CRLH-TL operates in the fast wave region ($v_p = \omega/\beta_x > c = \omega/k_0$), the propagating wave will be coupled into the free-space radiation, with a real value of k_z. On the other hand, if k_z is imaginary, the wave will decay exponentially along the z-direction, which results in an evanescent mode of the field.

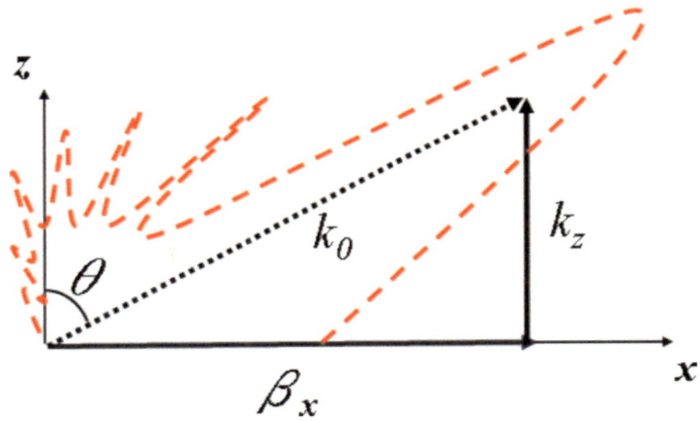

Figure 5.
Illustration of radiation phenomenon.

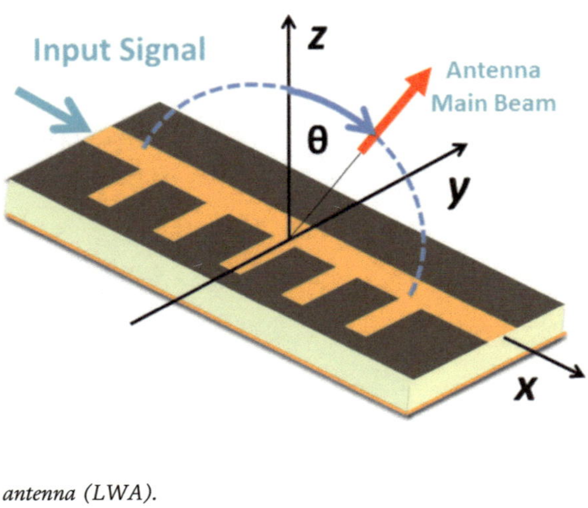

Figure 6.
1D CRLH leaky wave antenna (LWA).

From the vectorial relation shown in **Figure 5**, the direction of the main beam of radiation can be determined as

$$\theta_{MB} = \sin^{-1}\left(\frac{\beta_x}{k_0}\right) \tag{9}$$

Eq. (9) is approximately valid in many practical antennas, provided that $\alpha \ll \beta_x$. It can be known from **Figure 6** that since β_x is a function of frequency, the angle θ also varies with respect to the frequency, thereby enabling the frequency-dependent beam scanning. Such property is useful in many wireless systems that require the beam scanning function, as it does not need complex feeding networks and phase shifters used in conventional phased arrays. Eq. (9) also indicates that an ideal backfire-to-endfire (−90° to +90°) radiation may be possible, provided that the propagation constant is properly engineered.

3. Free space scanning and adaptative CRLH LWAS

As discussed above, CRLH LWAs can exhibit frequency-dependent beam scanning capabilities for continuously varying the beam from backfire to endfire directions by frequency tuning. Using this unique property, novel tunable and adaptive LWAs can be realized [20–28]. This section will discuss the practice of MTM-based LWAs and the experimental demonstrations for free space scanning capability with enhanced functionality.

The radiation pattern of 1D CRLH LWA can be represented using the array factor approach, assuming the antenna is oriented along x-direction [18]:

$$AF(\theta) = \sum_{n=1}^{N} I_n e^{j(n-1)k_0 p_x \sin\theta - j\varphi_{xn}} \tag{10}$$

where N is the number of the CRLH unit cells, p_x is the periodicity of unit cells, $\varphi_{xn} = (n-1)k_0 p_x \sin\theta_{MB}$, $I_n = I_0 e^{-\alpha(n-1)p_x}$ is an exponentially decaying function determined by the leakage factor α. As β for CRLH TLs can vary from $-k_0$ to $+k_0$, the main beam of 1D CRLH LWA can scan continuously from θ = −90° to +90° ($-x$ to $+x$ direction) as indicated in (9).

A typical microstrip realization for a CRLH-TL using the printed circuit board (PCB) technology is shown in **Figure 7** [8]. The unit cell of the CRLH-TL contains an interdigital capacitor and a stub inductor to form the desired LH elements. The size of unit cell is much smaller than the guided wavelength such that the structure behaves as an effectively homogenized in the frequency range of interest. Assuming

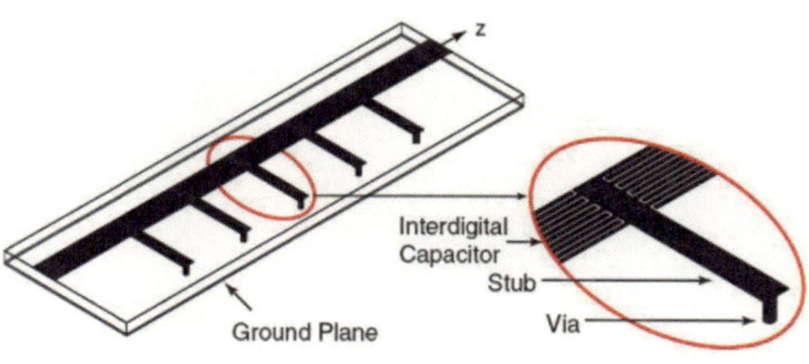

Figure 7.
Microstrip implementation of for a 1D CRLH-TL MTMstructure comprising periodically repeated sub-wavelength unit cells [8].

a lossless scenario, the unit cell model of this planar MTM TL can be described by the equivalent circuit in **Figure 3**. Owing to the inevitable parasitic effects, the right-handed series inductance and shunt capacitance need to be included, which forms an entire CRLH unit cell exhibiting a modal dispersion as shown in **Figure 4**: LH components dominate at lower frequencies, whereas RH components dominate at higher frequencies. A more rigorous unit cell model should also take into account the radiation loss (leakage rate), which are associated with resistance R and conductance G in the TL circuit model [15]. It is worth mentioning that the characteristic impedance is nearly independent of frequency for both RH and LH operations if properly designed, allowing a broadband impedance matching (i.e., 50 Ω in most RF and microwave systems).

Figure 8 shows a 24-cell CRLH LWA prototype and the measured radiation pattern. It is evident that as the frequency increases, the main beam direction scans from backward to forward, achieving a one-dimensional frequency scanning [5]. Ideally, a CRLH LWA scans a broad range of spatial region from $-90°$ to $+90°$, as frequency varies from $\omega = \beta \cdot c$ to $\omega = +\beta \cdot c$, including the broadside direction ($\theta = 0°$), which is usually at the evanescent mode for conventional LWAs [17].

In order to perform 2D frequency scanning, the antenna array must scan along another orthogonal direction, that is, y-direction as well. This can be achieved by properly designing the delay lines to manipulate the phase responses of each

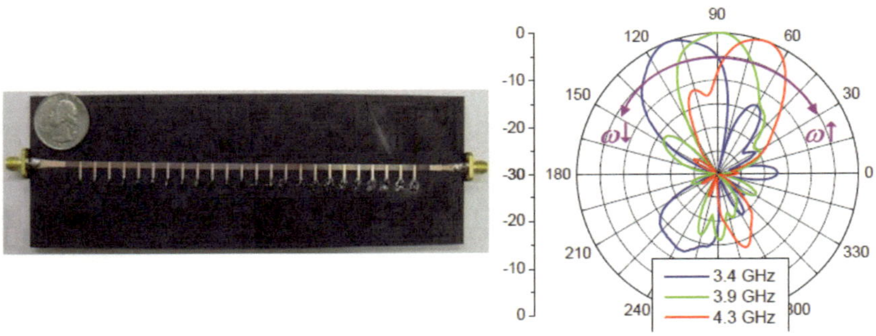

Figure 8.
Prototype of **Figure 7** *and its measured radiation patterns for different operating frequencies [8].*

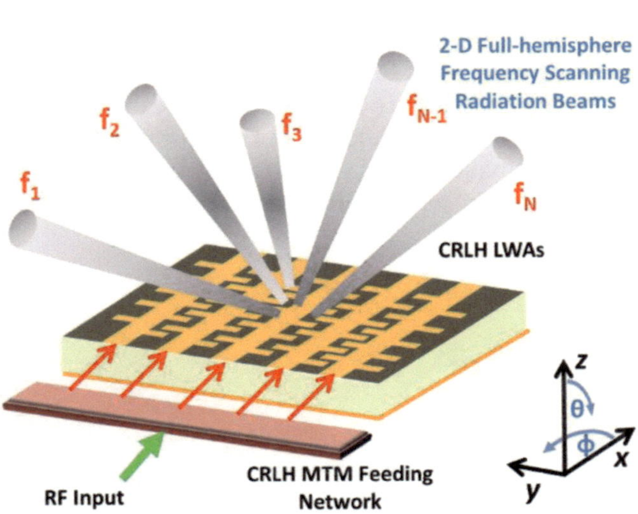

Figure 9.
Schematic of the proposed 2D MTM frequency scanning array exhibiting one-to-one frequency-space mapping [29].

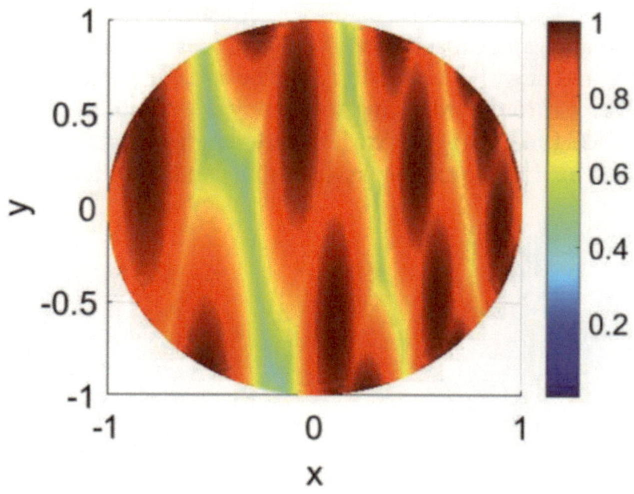

Figure 10.
Theoretical radiation beams using 2D array factor approach in (13) with L = 32, M = 2, N = 20. The frequency sweeps from 3 to 6.2 GHz, with an interval of 200 MHz, while each of the main beams corresponds to a distinct frequency [29].

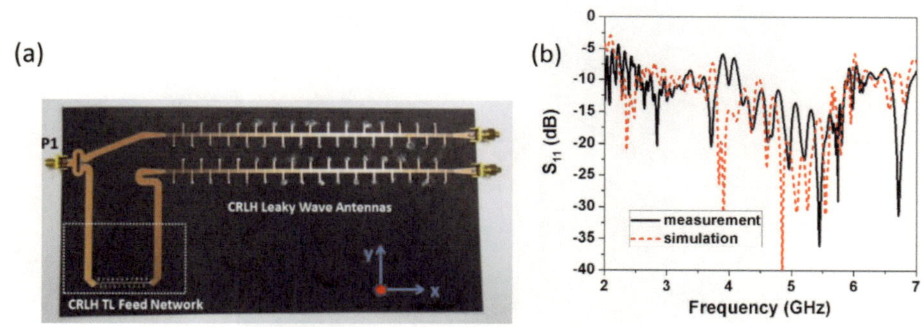

Figure 11.
Prototype of the proposed 2D MTM frequency scanning array. Measured and simulated return loss of the antenna array [29].

antenna element. To illustrate, the main beam angle of a 1D phased array is a function of progressive phase shift ξ, and can be represented as [19, 30]:

$$\theta_{array}(\omega) = \sin^{-1}\left(\frac{\xi}{k_0 d}\right) \tag{11}$$

Nevertheless, conventional delay lines can only provide phase delay, which bounds the radiated beam angles to only half of the hemisphere. To this end, if CRLH-TLs are used as feeding structures, they are able to provide both phase advance and phase delay, while exhibiting a zero phase-shift at the center operating frequency. Combining the CRLH-based feed network with 1D CRLH LWAs, the resulting 2D CRLH array can perform a 2D frequency-space mapping scheme as illustrated in **Figure 9** [29].

To explain, the phase response of a balanced CRLH TL consisting of N cascaded unit cells can be expressed as

$$\phi_{CRLH}(\omega) = -N\omega\sqrt{L_R C_R} + \frac{N}{\omega\sqrt{L_L C_L}} \tag{12}$$

where L_R, C_R and L_L, C_L are right-handed inductance and capacitance, as well as left-handed inductance and capacitance of a CRLH unit cell. It is observed from (12) that both positive and negative phase delay can be achieved using the CRLH feeding structures with respect to the operating frequency. Furthermore, if the phase response of the CRLH feed network varies fast enough within each sub-band of CRLH LWAs, a sequential 2D frequency scanning can be realized.

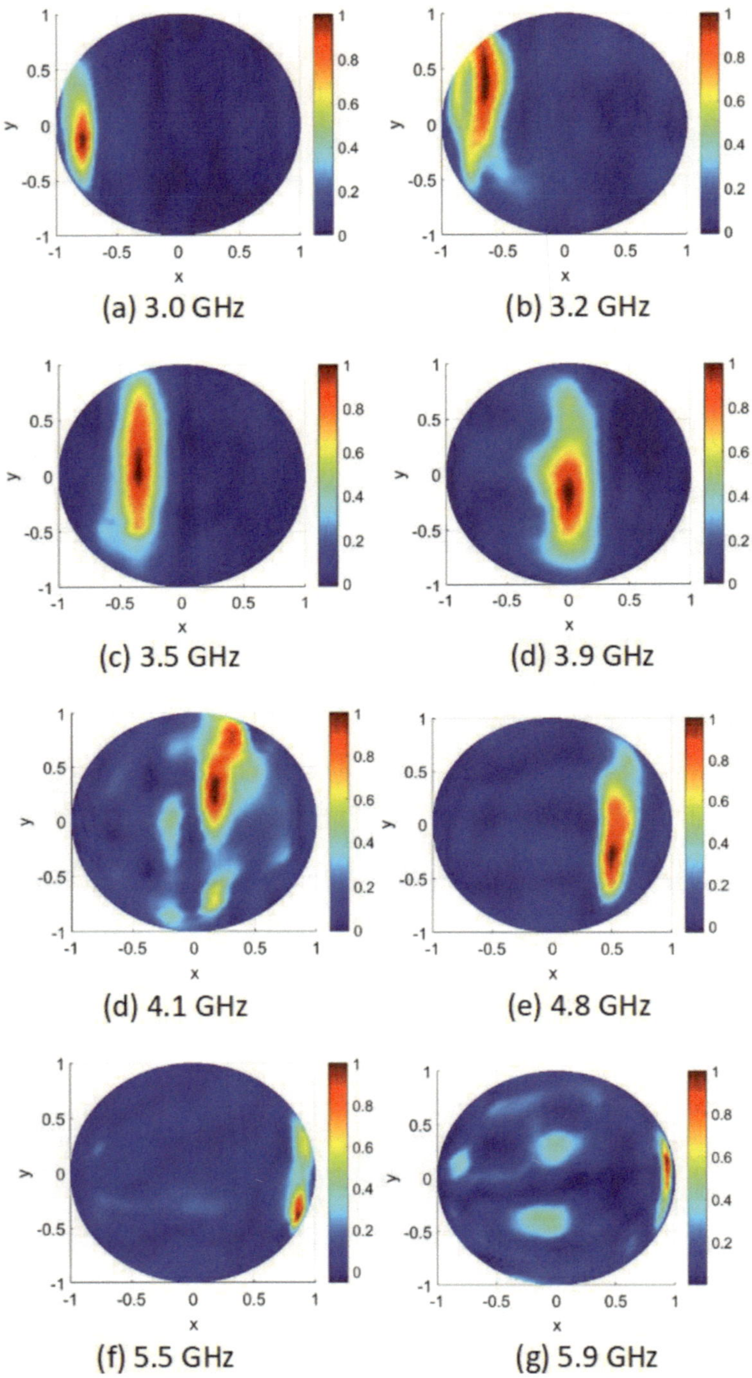

Figure 12.
Measured radiation patterns of the proposed 2D MTM frequency scanning array [29].

The 2D frequency scanning CRLH LWA array can be analyzed using the theoretical model based on 2D array factor approach. Its mathematical form is represented by the multiplication of the array factor of a 1D CRLH LWA and the one resulted from the phased array fed by the CRLH feed networks:

$$AF(\theta,\phi) = \sum_{n=1}^{L} I_n e^{j(n-1)k_0 p_x \sin\theta\cos\phi - j\varphi_{xn}} \times \sum_{m=1}^{M} e^{j(m-1)k_0 p_y \sin\theta\sin\phi + j\varphi_{yn}} \tag{13}$$

where L is the number of the unit cells for a CRLH LWA, p_x is the periodicity of LWA unit cells, $\varphi_{xn} = (n-1)k_0 p_x \sin\theta_{MB}$, $I_n = I_0 e^{-\alpha(n-1)p_x}$ is an exponentially decaying function determined by the leakage factor α. Moreover, p_y is the spacing between the CRLH LWAs, M is the number of LWAs and $\varphi_{yn} = (m-1)\left(-N\omega\sqrt{L_R C_R} + \frac{N}{\omega\sqrt{L_L C_L}}\right)$, which is the phase response of CRLH feed lines with N unit cells.

For proof-of-concept, the CRLH LWA is designed to operate at 3–6.2 GHz. **Figure 10** plots the results of 2D MTM array obtained from (13) with a 200 MHz sweeping interval. For simplicity, the spatial location and radiation intensity are both normalized, where $x = \sin\theta\cos\phi, y = \sin\theta\sin\phi$. As a result, the resulting 2D CRLH aperture exhibit one-to-one frequency-space mapping characteristics covering the entire hemisphere above the radiating aperture, with each spectral component pointing to a specific angle in the free space.

To verify, a prototype of 2D MTM LWA array using printed circuit board (PCB) technology is shown in **Figure 11a**. The antenna array prototype contains two CRLH LWAs and a lumped component-based CRLH feed network embedded in a Wilkinson power divider. The return loss of the proposed structure is shown in **Figure 11b**. The measured results agree reasonably well with the EM simulation using ANSYS Electronics, indicating a good return loss within the operating band of 3–6.2 GHz. The slight mismatch at the center frequency around 4 GHz might be due to the handmade fabrication errors.

The measured radiation patterns of various frequencies within the antenna operating band are plotted in **Figure 12**. As can be seen, the main beam of the 2D

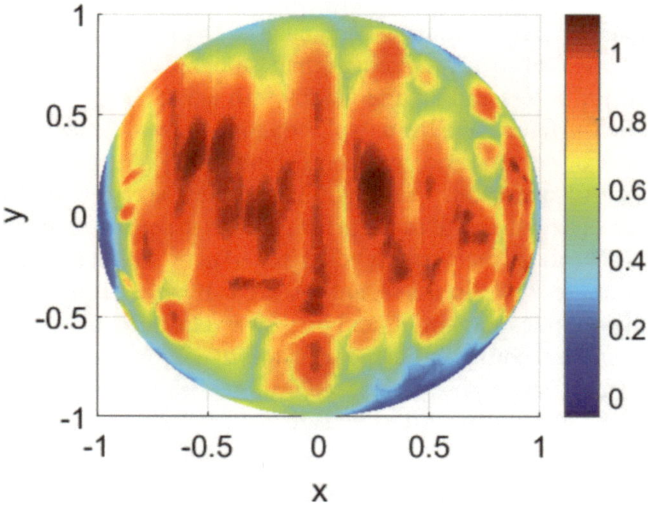

Figure 13.
Measured overlaid radiation beams of the fabricated 2D MTM array sweeping from 3 to 6.2 GHz with 25 MHz frequency steps [29].

MTM array is able to sweep in two dimensions, namely x-direction and y-direction. The x-direction scanning is due to the nature of CRLH LWAs, whereas the y-direction scanning results from the CRLH feed network that provides the required progressive phases. This mechanism leads to 2D frequency-to-space mapping capability. In addition, **Figure 13** overlays all the radiation patterns of the entire operating band from 3 to 6.2 GHz with an interval of 25 MHz. It can be observed that the main beams are frequency-mapped to the entire 2D full-hemisphere space.

4. Beamforming based on planar metasurfaces

This section discusses a new type of holographic antenna using the ultralow-profile metasurface hologram. Similar to the concept of transformation optics using MTMs, radiation from the metasurface relies on transforming the guided surface wave into the free-space radiation leakage, through its interactions with the anisotropic reactance tensor of metasurface defined by the analytical formula. The metasurface with an inhomogeneous and anisotropic surface impedance profile can be implemented using dense, textured subwavelength metallic structures printed on a printed circuit board. The beam angle with respect to the propagation direction of the surface wave depends on the spatial profile of surface impedance of the metasurface. Like the TL-MTM antenna, the metasurface antennas can be built with the standard PCB process and be excited by a simple in-plane feeder or a plane wave excitation. Therefore, the metasurface-based holographic antennas show significant potential for making compact, low-profile and conformal surface-like antennas, which have capabilities of generating a high-gain and steerable beam for satellite and space applications. Unlike the transformation optics approaches based on bulky MTMs [27, 28, 31, 32], where the control on propagation often requires highly anisotropic values for $\overline{\overline{\varepsilon}}$ and $\overline{\overline{\mu}}$ tesnors, the applicability of transformation techniques to 2D metasurface can avoid the complex transformation structures, power losses and manufacturing cost.

A metasurface can be seen as a 2D version of metamaterial, which is formed by arranging subwavelength scatters or apertures into a surface or interface [33–35]. Through the design of size, period and shape of textured surface, the metasurface can display a quasi-homogeneous or anisotropic surface impedance profile [34–38]. For many applications, metasurfaces have been be used in place of bulky MTMs because they occupy less physical space, reducing remarkably the manufacturing cost, and providing lower power loss and broader bandwidth when compared with their MTM counterparts. Metasurfaces have been exploited to make novel electromagnetic devices with superior performance and compactness [33]. The metasurface concept has emerged as an advantageous reconfigurable antenna architecture for beam forming and wave-front shaping. Metasurface antennas consist of an array of subwavelength inclusions (meta-atoms) distributed over an electrically large structure. In this section, we will discuss recent research advances in metasurfaces, which bring new exciting applications in RF and microwave beamforming technologies. Particularly, using the reconfigurable metasurface in conjunction with the holography principle enables new types of holographic antennas that electronically steer the beam, with a high realized gain and a low sidelobe level.

4.1 Principle of holographic antennas

Holographic antennas are a class of antennas whose radiation apertures are formed by the discrete spatially distributed diffraction apertures. As inspired by the optical holography, the surface impedance profile of such an inhomogeneous planar aperture can be generated by the interference pattern of the desired (signal) beam with that of the reference beam. The concept of microwave holography or holographic antenna was first demonstrated by Checcacci, Russo and Scheggi in 1970 [39]. Microwave holographic antennas have been widely investigated in the past decades [40]. Typically, microwave holographic antennas typically comprise a main surface-wave-carrying impedance surface, which is separated from the ground plane by a dielectric slab. Such an impedance surface can be designed to serve as a guiding structure or to support leaky waves, depending on the eigenmodes of the surface-wave structure and/or the modulation period and depth of equivalent surface impedance [39–46]. The pattern of surface perturbations and spatial profile of surface impedance can be designed by a "holography" method [41]. In this context, the electrically thin metasurface has been known for its capabilities of controlling the phase and amplitude fronts of surface waves [40–48]. As a result, metasurfaces may serve as an ideal platform for microwave holograms that gives control to the amplitude, phase and polarization of aperture fields.

The holography process involves producing an interference pattern using two waves, and then using the interference pattern to scatter one wave to produce the other. The interference pattern formed by these two waves is recorded as a *hologram*. When the reference wave illuminates the hologram, it is scattered by the recorded interference pattern to produce a copy of the original object wave. For a reference wave Ψ_{ref} and an object wave Ψ_{obj}, the interference pattern contains a term proportional to $\Psi_{obj}\Psi_{ref}^*$. When the interference pattern is illuminated by the reference wave, it renders $\left(\Psi_{obj}\Psi_{ref}^*\right)\Psi_{ref} = \Psi_{obj}\left|\Psi_{ref}\right|^2$, forming a copy of the original object wave. For the case of a leaky-wave antenna excited by the bound transverse magnetic (TM) surface modes, we may define the geometrical surface perturbation as the interference pattern between these two waves [41]:

$$Z(x,y) = j\left|X + M\ \mathrm{Re}\left[\Psi_{rad}\Psi_{ref}^*\right]\right|, \tag{14}$$

where X and M are the arbitrary real average value and the real modulation depth at the position (x,y) on the impedance surface; we assume the surface lies in the x-y plane. The radiated (object) wave can be seen as the product of surface-wave scattering from a modulated surface, given by $\Psi_{obj}\left|\Psi_{ref}\right|^2$.

For the one-dimensional applications (**Figure 14**), the hologram is produced by the interference pattern of a 1D surface wave and a plane wave propagating in free space at an θ_0 Sampling the hologram impedance function at each unit cell center gives the local impedance value and hence the geometry and size of local metal patch.

In principle, generation of arbitrary polarization requires an anisotropic and inhomogeneous metasurface. Consider a uniform and anisotropic metasurface lies on the x-y plane, the averaged tangential electric field at the surface $\overline{\mathbf{E}}\big|_{z=0^+}$ is related to the averaged induced surface current $\overline{\mathbf{J}}_{\mathbf{s}}$ by $\overline{\mathbf{E}}\big|_{z=0^+} = \overline{\overline{\mathbf{Z}}}\cdot\left(\hat{z}\times\overline{\mathbf{H}}\right)\big|_{z=0^+} = \overline{\overline{\mathbf{Z}}}\cdot\overline{\mathbf{J}}_{\mathbf{s}}$, where the surface impedance dyad $\overline{\overline{\mathbf{Z}}}$ appearing in a tensor form is given by [41, 42]:

$$\begin{pmatrix} E_x \\ E_y \end{pmatrix} = \begin{pmatrix} Z_{xx} & Z_{xy} \\ Z_{yx} & Z_{yy} \end{pmatrix}\begin{pmatrix} J_x \\ J_y \end{pmatrix} \tag{15}$$

In the lossless scenario, the surface impedance tensor $\overline{\overline{\mathbf{Z}}}$ is anti-Hermitian (i.e., $\overline{\overline{\mathbf{Z}}} = -\overline{\overline{\mathbf{Z}}}*T$), and the applicability of reciprocity enforces that $\overline{\overline{\mathbf{Z}}}$ is purely imaginary

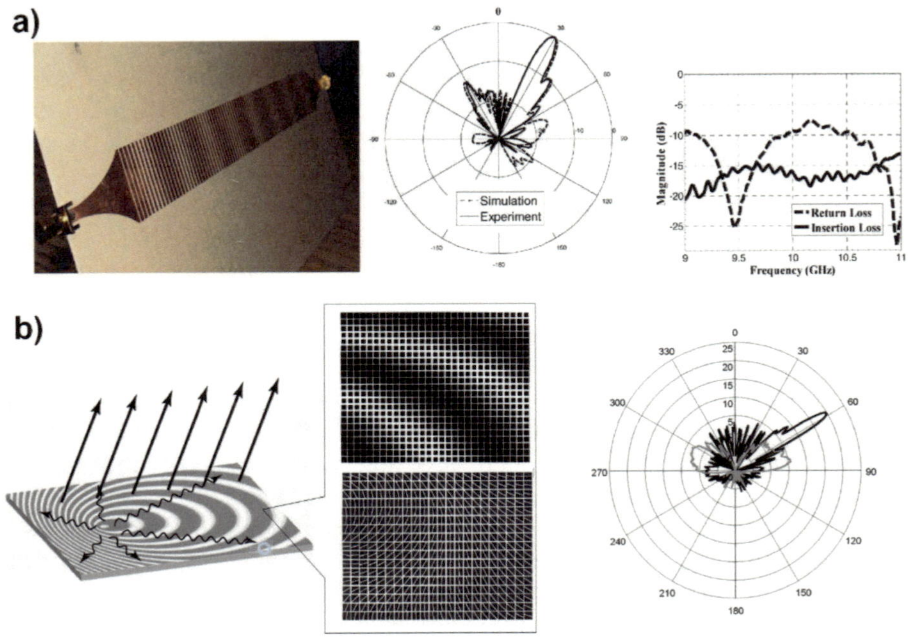

Figure 14.
(a) Left: 1D holographic leaky wave antenna using a metasurface composed of a subwavelength metal strips on top of a grounded dielectric substrate. The local surface reactance is varied by adjusting the strip width. Right: Measured radiation patterns and bandwidth of operation [43]. (b) 2D holographic leaky wave antenna using a metasurface and a zoomed-in hologram portion and unit cells (metallic patch of different geometric shapes that are printed on top of a grounded dielectric substrate). The undulating and straight arrows represent the surface wave and desired radiation, respectively. The unit cell size is fixed at 3 mm, which is much smaller than the free space wavelength. Measured radiation patterns from a 3 mm monopole antenna placed above the holographic metasurface (black) and an untextured metal surface (gray) at 17 GHz. The inclination angle is defined from the impedance surface normal, that is, the z-axis in the x-z plane [42].

and $Z_{xy} = Z_{yx}$. The surface impedance tensor is constructed from the outer product of the expected surface current vectors and the desired outgoing electric field vectors. The outer product generalizes the simple multiplicative scalar pattern described earlier in this section. In the tensor case, the desired radiated vector wave term $\overline{\mathbf{E}}_{rad}\left|\overline{\mathbf{J}}_{surf}\right|^2$ is obtained from scattering of a vector surface wave from a modulated anisotropic impedance surface with $\overline{\overline{\mathbf{Z}}}$. When the anisotropic metasurface hologram is illuminated with the reference surface wave, a modulated tensor impedance proportional to $\overline{\mathbf{E}}_{rad} \otimes \overline{\mathbf{J}}_{surf}^{\dagger}$ can create a radiated vector wave term: $\overline{\overline{\mathbf{Z}}} \cdot \overline{\mathbf{J}}_{surf} \propto \overline{\mathbf{E}}_{rad} \otimes \overline{\mathbf{J}}_{surf}^{\dagger} \cdot \overline{\mathbf{J}}_{surf} = \overline{\mathbf{E}}_{rad} \cdot \left|\overline{\mathbf{J}}_{surf}\right|^2$ (the dagger represents the Hermitian conjugate). For a power-conserving and reciprocal metasurface, $\overline{\overline{\mathbf{Z}}}$ must be made anti-Hermitian and pure imaginary, given by:

$$\overline{\overline{\mathbf{Z}}} = j\begin{pmatrix} X & 0 \\ 0 & X \end{pmatrix} + j\frac{M}{2}\mathrm{Im}\left[\overline{\mathbf{E}}_{rad} \otimes \overline{\mathbf{J}}_{surf}^{\dagger} - \overline{\mathbf{J}}_{surf} \otimes \overline{\mathbf{E}}_{rad}^{\dagger}\right] \quad (16)$$

where the diagonal X in the impedance matrix binds the TM surface wave, and the modulated impedance components M scatter the vector surface wave into the desired vector outgoing wave. A circularly polarized planar antenna using the anisotropic holographic metasurface have been demonstrated in Ref. [44], of which the circularly polarized radiation is produced from a linearly polarized source through interactions with the transformative metasurface hologram.

The metasurface-based holographic antenna can be made extremely flat, light weight (<1 kg), low cost and built with the standard PCB processes. Since each elementary inclusion on the metasurface is subwavelength and non-resonant, this holographic antenna exhibits satisfactory insertion and return losses.

With the advent in the metasurface technology, one already has the ability to accurately define the local surface impedance profile of metasurface and conduct the inverse design of the spatially invariant inclusions over the surface. Several approaches have been proposed to design and characterize the surface impedance dyad of metasurfaces with great accuracy [48]. **Figure 15** shows a prototype of circularly polarized holographic antenna based on an anisotropic metasurface [44]. To produce the anisotropic impedance, the geometry of metallic patch must be asymmetric with respect to the propagation direction of the surface wave. The measured realized gain in the horizontal plane at 8.6 GHz is also presented in **Figure 15**. It is clearly seen that the metasurface-based holographic antenna exhibits an excellent contrast between the right-handed circular-polarization radiation (co-polarization) and left-handed circular-polarization radiation (cross-polarization) radiation. Noticeably, a broadband and broad-angle operation could be achieved with this compact metasurface antenna.

Combining the holography concept with the reconfigurable metasurface will further allow dynamic beamsteering capabilities, without the need of mechanical moving parts [49–57]. In 2013, the Kymeta Corporation has demonstrated an active metasurface antenna used for bi-directional high-speed internet connectivity [49]. In their design, the reconfigurability is achieved by using inclusions (i.e., meta-atoms) that can be individually tuned to create a dynamic hologram. In their design, controlling local radiative responses is possible by loading each meta-atom with liquid crystals that present a variable permittivity tensor. Moreover, a dynamic hologram with faster response time and lower loss may be achieved by exploiting meta-atoms loaded with active electronic components (e.g., transistors or diodes).

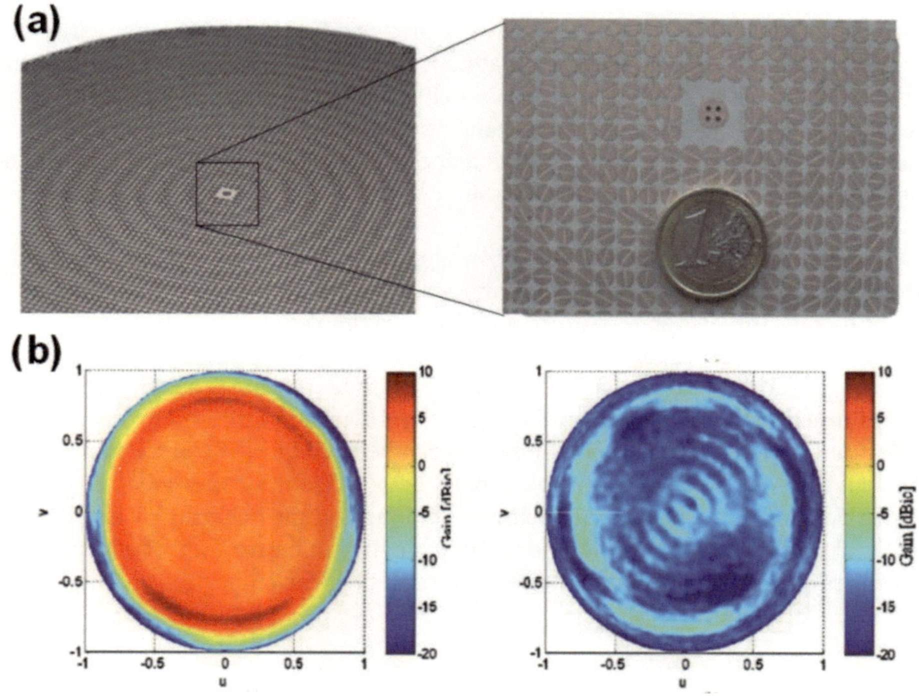

Figure 15.
(a) Photograph of the prototype of circularly polarized metasurface antenna and details of its feeding zone without exciter pin. (b) Gain measurements for the antenna at 8.6 GHz on the horizontal plane: (left) RHCP component and (right) LHCP component [44].

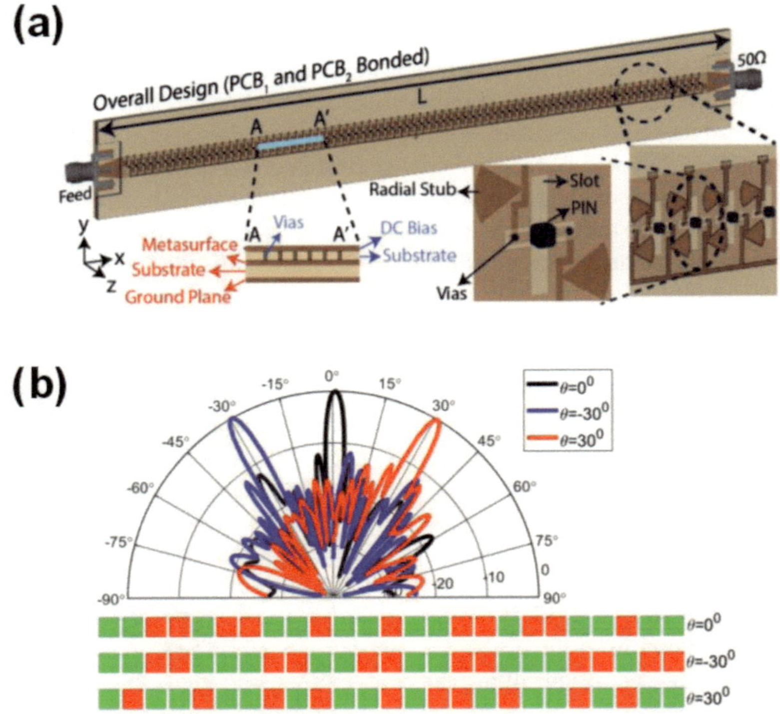

Figure 16.
Reconfigurable metasurface made of arrays of meta-atom apertures loaded with PIN diodes [51]. A tunable metasurface antenna: the input carrier wave traverses the waveguide and the meta-atom couple energy out of the feed; an unit cell is switched to its on- and off-states by the external circuitry. (b) When this holographic antenna is illuminated by an underlying feed wave, a coherent beam is produced. Different images, encoded in the array by the control circuit, produce various radiations over a broad angular range.

Ideally, each meta-atom must be weakly coupled and be capable of being individually driven to obtain different electromagnetic properties at the scale of a single unit cell. Enhancement of bandwidth is also possible by combining metasurfaces with non-Foster elements. Different from conventional phase arrays and leaky-wave antennas, a dynamic metasurface antenna does not require active phase shifters and amplifiers, and can achieve reconfigurability locked to specific operating frequencies variant from 8.5 to 8.8 GHz.

Figure 16 shows schematically a reconfigurable holographic antenna based on the active metasurface, of which an individual unit cell is loaded with active component to vary the hologram pattern and thus change angular distributions of radiation. Modeling the collective responses from meta-atoms and the subsequent radiation pattern have been studied using the array factor method and the Dyadic Green's functions for dipole excitations [53, 56]. In addition, the analytical model for design of meta-atoms of different kinds can be found in [54]. Since each meta-atom can be modeled as a polarizable dipole, its polarizability and radiative properties can be well defined. These modeling technique and the holographic explanation shed light on important considerations for achieving an better control of radiation patterns produced by the active metasurface antenna [50–55].

5. Conclusions

Planar MTM technologies manipulating the electromagnetic waves, somehow similar to the transformation optics, may suggest new types of efficient beamforming and beamshaping techniques for the modern wireless systems. This chapter has presented some of the most recent and representative innovations in the planar TL-MTMor metasurface antenna based on the microstrip implementation. Their operation principles basically rely on tailoring the propagation (phase) constant and impedance by using the homogenized, subwavelength reactive elements in an effective medium. Further, the tunability in phase velocity, propagation constant and impedance can be achieved with the combinational responses associated with the periodic nature of metamaterial/metasurface structure and with the loaded lumped components (i.e., diodes, varactors, switches and transistors) in the unit cell. This enables the beamforming and beamsteering antennas to be made thinner, lighter weight, less expensive and requiring less power and having enhanced bandwidth than the conventional alternatives; for instance, phase arrays typically include multiple active antennas with costly phase shifters, amplifiers and complex feeding structures. The novel surface antenna technologies, being capable of dynamically transforming the guided waves to the desired radiation patterns, show great potential to serve as beamformers and beamsteerers for the next-generation wireless technology [58, 59].

Author details

Chung-Tse Michael Wu[1*] and Pai-Yen Chen[2]

1 Department of Electrical and Computer Engineering, Rutgers, The State University of New Jersey, NJ, USA

2 Department of Electrical and Computer Engineering, University of Illinois at Chicago, Chicago, IL, USA

*Address all correspondence to: ctm.wu@rutgers.edu

References

[1] Sihvola A. Metamaterials in electromagnetics. Meta. 2007;**1**(1):2-11

[2] Smith DR, Padilla WJ, et al. Composite medium with simultaneously negative permeability and permittivity. Physical Review Letters. 2000;**84**(18): 4184-4187

[3] Shelby RA, Smith DR, et al. Experimental verification of a negative index of refraction. Science. 2001; **292**(5514):77-79

[4] Yen TJ, Padilla WJ, et al. Terahertz magnetic response from artificial materials. Science. 2004;**303**(5663):1494

[5] Caloz C, Itoh T. Electromagnetic Metamaterials: Transmission Line Theory and Microwave Applications. Hoboken, New Jersey: John Wiley & Sons; 2005

[6] Eleftheriades GV, Iyer AK, Kremer PC. Planar negative refractive index media using periodically LC loaded transmission lines. IEEE Transactions on Microwave Theory and Techniques. 2002;**50**(12):2702-2712

[7] Engheta N, Ziolkowski RW, editors. Metamaterials: Physics and Engineering Explorations. Hoboken, New Jersey: John Wiley & Sons; 2006

[8] Lai A, Itoh T, Caloz C. Composite right/left-handed transmission line metamaterials. IEEE Microwave Magazine. 2004;**5**(3):34-50

[9] Kuylenstierna D, Vorobiev A, Linner P, Gevorgian S. Composite right/left handed transmission line phase shifter using ferroelectric varactors. IEEE Microwave and Wireless Components Letters. 2006;**16**(4):167-169

[10] Tseng CH, Itoh T. IEEE MTT-S International Microwave Symposium Digest. IEEE; 2006. pp. 931-934

[11] Eleftheriades GV, Grbic A, Antoniades M. Negative-refractive-index transmission-line metamaterials and enabling electromagnetic applications. In: IEEE APS-S International Antennas and Propagation Society International Symposium. IEEE; 2004. pp. 1399-1402

[12] Eleftheriades GV. EM transmission line metamaterials. Materials Today. 2009;**12**(3):30-41

[13] Liu L, Caloz C, Itoh T. Dominant mode leaky-wave antenna with backfire-to-endfire scanning capability. Electronics Letters. 2002;**38**(23): 1414-1416

[14] Allen CA, Leong KM, Itoh T. 2-D frequency-controlled beam-steering by a leaky/guided-wave transmission line array. In: IEEE MTT-S International Microwave Symposium Digest. IEEE; June 2006. pp. 457-460

[15] Lim S, Caloz C, Itoh T. Metamaterial-based electronically controlled transmission-line structure as a novel leaky-wave antenna with tunable radiation angle and beamwidth. IEEE Transactions on Microwave Theory and Techniques. 2004;**52**(12): 2678-2690

[16] Casares-Miranda FP, Camacho-Peñalosa C, Caloz C. High-gain active composite right/left-handed leaky-wave antenna. IEEE Transactions on Antennas and Propagation. 2006;**54**(8): 2292-2300

[17] Oliner AA. In: John RC, editor. Leaky-Wave Antennas. Antenna Engineering Handbook. 3rd ed. New York, New York: McGraw-Hill; 1993

[18] Caloz C, Itoh T. Array factor approach of leaky-wave antennas and application to 1-D/2-D composite right/left-handed (CRLH) structures.

IEEE Microwave and Wireless Components Letters. 2004;**14**(6): 274-276

[19] Choi J, Sun JS, Itoh T. Frequency-scanning phased-array feed network based on composite right/left-handed transmission lines. IEEE Transactions on Microwave Theory and Techniques. 2013;**61**(8):3148-3157

[20] Huang L, Chiao J, Lisio P. An electronically switchable leaky wave antenna. IEEE Transactions on Antennas and Propagation. 2000; **48**(11):1769-1772

[21] Maheri H, Tsutsumi M, Kumagi N. Experimental studies of magnetically scannable leaky-wave antennas having a corrugated ferrite slab/dielectric layer structure. IEEE Transactions on Antennas and Propagation. 1988;**36**(7): 911-917

[22] Wu C-TM, Dong Y, Sun JS, Itoh T. Ring-resonator-inspired power recycling scheme for gain-enhanced distributed amplifier-based CRLH-transmission line leaky wave antennas. IEEE Transactions on Microwave Theory and Techniques. 2012;**60**(4): 1027-1037

[23] Wu C-TM, Itoh T. Combined gain-enhanced power recycling feedbacks for distributed amplifier-based crlh-leaky wave antennas. In: 41st European Microwave Conference (EuMC). IEEE; 2011. pp. 499-502

[24] Wu C-TM, Itoh T. Gain-enhanced distributed amplifier-based CRLH-leaky wave antenna for quasi-resonant power recycling scheme. In: Microwave Symposium Digest (MTT). IEEE; 2011. pp. 1-4

[25] Wu C-TM, Itoh T. Dual-fed distributed amplifier-based CRLH-leaky wave antenna for gain-enhanced power combining. In: IEEE MTT-S International Microwave Workshop Series on Innovative Wireless Power Transmission - Technologies, Systems, and Applications (IMWS). IEEE; 2012, May. pp. 87-90

[26] Wu C-TM, Itoh T. CRLH-transmission line leaky wave antennas integrated with distributed amplifiers with power recycling feedback scheme. In: Proceedings of the 5th European Conference on Antennas and Propagation (EUCAP). IEEE; April 2011. pp. 3901-3904

[27] Pendry JB, Schurig D, Smith DR. Controlling electromagnetic fields. Science. 2006;**312**(5781):1780-1782

[28] Leonhartdt U. Optical conformal mapping. Science. 2006;**312**(5781): 1777-1780

[29] Salarkaleji M, Ali MA, Wu C-TM. Two-dimensional full-hemisphere frequency scanning array based on metamaterial leaky wave antennas and feed networks. In: IEEE MTT-S International Microwave Symposium (IMS). IEEE; 2016. pp. 1-4

[30] Stutzman W, Thiele GA. Antenna Theory and Design. Hoboken, New Jersey: John Wiley & Sons; 2012

[31] Chen H, Chan CT, Sheng P. Transformation optics and metamaterial. Nature Materials. 2010;**9**: 387-396

[32] Ma HF, Cui TJ. Three-dimensional broadband and broad-angle transformation-optics lens. Nature Communications. 2010;**1**:124

[33] Holloway CL, Kuester EF, et al. An overview of the theory and applications of metasurfaces: The two-dimensional equivalents of metamaterials. IEEE Antennas and Propagation Magazine. 2012;**54**(2):10-35

[34] Holloway CL, Mohamed MA, et al. Reflection and transmission properties

of a metafilm: With an application to a controllable surface composed of resonant particles. IEEE Transactions on Antennas and Propagation. 2005;**47**(4): 853-865

[35] Holloway CL, Kabos P, et al. Realization of a controllable metafilm/ metasurface composed of resonant magnetodielectric particles: Measurements and theory. IET Microwaves, Antennas and Propagation. 2010;**4**(8):1111-1122

[36] Chen PY, Alù A. Mantle cloaking using thin patterned metasurfaces. Physical Review B. 2011;**84**(20):205110

[37] Padooru YR, Yakovlev AB, Chen PY, Alù A. Analytical modeling of conformal mantel cloaks for cylindrical objects using sub-wavelength printed and slotted arrays. Journal of Applied Physics. 2012;**112**(3):034907

[38] Padooru Y, Yakovlev R, Chen PY, Alù A. Line-source excitation of realistic conformal metasurface cloaks. Journal of Applied Physics. 2012;**112**(10): 104902

[39] Checcacci P, Russo V, Scheggi A. Holographic antennas. IEEE Transactions on Antennas and Propagation. 1970;**18**(6):811-813

[40] ElSherbiny M, Fathy AE, et al. Holographic antenna concept, analysis and parameters. IEEE Transactions on Antennas and Propagation. 2004;**52**(3): 830-839

[41] Fong BH, Colburn JS, et al. Scalar and tensor holographic artificial impedance surfaces. IEEE Transactions on Antennas and Propagation. 2010; **58**(10):3212-3221

[42] Sievenpiper D, Colburn J, et al. Holographic artificial impedance surfaces for conformal antennas. In: Antennas and Propagation Society International Symposium. IEEE; 2005. pp. 256-259

[43] Patel AM, Grbic A. A printed leaky-wave antenna based on a sinusoidally-modulated reactance surface. IEEE Transactions on Antennas and Propagation. 2011;**59**(6):2087-2096

[44] Minatti G, Maci S, et al. A circularly-polarized isoflux antenna based on anisotropic metasurface. IEEE Transactions on Antennas and Propagation. 2012;**60**(11):4998-5009

[45] Minatti G, Caminita F, et al. Spiral leaky-wave antennas based on modulated surface impedance. IEEE Transactions on Antennas and Propagation. 2011;**59**(12):4436-4444

[46] Chen PY, Farhat M, et al. Infrared beam-steering using acoustically modulated surface plasmons over a graphene monolayer. Journal of Optics. 2014;**16**(9):094008

[47] Eleftheriades GV, Wong AMH. Holography-inspired screens for sub-wavelength focusing in the near field. IEEE Microwave and Wireless Components Letters. 2008;**18**(4): 236-238

[48] Patel AM, Grbic A. Transformation electromagnetics devices using tensor impedance surface. In: MTT-S International Microwave Symposium Digest (IMS). IEEE; 2013. pp. 1-4

[49] Bleicher A. Kymeta Demos First Ever Satellite Link with Metamaterial Antenna. New York: IEEE Spectrum; 2013

[50] Johnson M, Bowen P, et al. Discrete-dipole approximation model for control and optimization of a holographic metamaterial antenna. Applied Optics. 2014;**53**(25):5791-5799

[51] Yurduseven O et al. Dynamically reconfigurable holographic metasurface aperture for a Mills-cross monochromatic microwave camera. Optics Express. 2018;**26**(5):5281-5291

[52] Yurduseven O et al. Millimeter-wave spotlight imager using dynamic holographic metasurface antennas. Optics Express. 2017;**25**(15): 18230-18249

[53] Smith DR et al. Analysis of a waveguide-fed metasurface antenna. Physical Review Applied. 2017;**8**(5): 054048

[54] Pulido-Mancera L et al. Polarizability extraction of complementary metamaterial elements in waveguides for aperture modeling. Physical Review B. 2017;**96**(23):235402

[55] Johnson M, Brunton SL, et al. Sidelobe canceling on a reconfigurable holographic metamaterial antenna. In: International Conference on Electromagnetics in Advanced Applications (ICEAA). 2014. pp. 806-809

[56] Lipworth G, Mrozack A, et al. Metamaterial apertures for coherent computational imaging on the physical layer. Journal of the Optical Society of America A. 2013;**30**(8):1603-1612

[57] Hunt J, Driscoll T, et al. Metamaterial apertures for computational imaging. Science. 2013; **339**(6117):310-313

[58] Mori K, Itoh T. Distributed amplifier with CRLH-transmission line leaky wave antenna. In: EuMC 2008. 38th European Microwave Conference. IEEE; 2008. pp. 686-689

[59] Wu C-TM, Itoh T. A re-radiating CRLH-transmission line leaky wave antenna using distributed amplifiers. In: APMC 2009. Asia Pacific Microwave Conference. IEEE; 2009. pp. 1998-2001

Chapter 3

High Performance Metasurface Antennas

Haisheng Hou, Haipeng Li, Guangming Wang, Tong Cai, Xiangjun Gao and Wenlong Guo

Abstract

Recently, metasurfaces (MSs) have received tremendous attention because their electromagnetic properties can be controlled at will. Generally, metasurface with hyperbolic phase distributions, namely, focusing metasurface, can be used to design high-gain antennas. Besides, metasurface has the ability of controlling the polarization state of electromagnetic wave. In this chapter, we first propose a new ultrathin broadband reflected MS and take it into application for high-gain planar antenna. Then, we propose multilayer multifunctional transmitted MSs to simultaneously enhance the gain and transform the linear polarization to circular polarization of the patch antenna. This kind of high-gain antenna eliminates the feed-block effect of the reflected ones.

Keywords: focusing, metasurface, high-gain, polarization conversion, reflection, transmission

1. Single-layer broadband planar antenna using ultrathin high-efficiency focusing metasurfaces

Recently, metasurfaces (MSs) have attracted growing interests of many researchers due to their planar profile, easy fabrication, and also strong beam control capacity [1–6]. For phase gradient metasurfaces (PGMS), proposed by Yu et al. [7], a wide range of applications have been found, such as anomalous beam bending [8, 9], focusing [7, 10], surface-plasmon-polariton coupling [11, 12], and polarization manipulation. With the ability of tuning the phase range covering 2π, metasurface can be used to improve performance of antenna. By fixing proper phase distributions on the metasurface, we can manipulate the wavefronts and the polarizations of the electromagnetic waves (EM) at will.

The focusing metasurface, which is one kind of functional metasurfaces, can focus the incident plane wave to its focal point. These characters indicate that the focusing metasurface can be applied for designing planar high-gain antenna by placing the feed sources at the focal point over the focusing metasurface [1]. Generally, there are two types of focusing metasurface, namely, reflective focusing metasurface and transmitted focusing metasurface. Based on upon two types of focusing metasurface, there are two kinds of high-gain antenna, namely, reflective metasurface high-gain antenna and transmissive metasurface high-gain antenna. Compared with the reflective metasurface antenna based on the focusing metasurface, transmissive metasurface antenna avoids the feed blockage effect, making it more suitable for a high-gain antenna design. Besides, when designing a reflective focusing metasurface, the reflection magnitude is close to 1 (0 dB) due to grounded plane composed of PEC. Therefore, the reflection phase is the only modulated character by the focusing metasurface. However, transmitted phase and amplitude are both needed to be considered when designing a transmissive focusing metasurface. In [1], a dual-mode and dual-band flat high-gain antenna based on focusing metasurface is proposed. The reflection beam and transmission beam can be obtained. In [2], the modified I-shaped particles, which can independently manipulate the phases and amplitudes of the cross-polarization waves, have been proposed. Based on the proposed unit cell, three high-gain antennas are fabricated and tested. In [5], a novel split beam antenna using transmission-type coding metasurface is proposed, which provides a new way to design focusing metasurface. Although reflective metasurface antenna and transmissive metasurface antenna have been studied for many years, more efforts should be done to improve the performance (such as realized gain, bandwidth, polarization states, and so on) of antenna.

However, most reported metasurfaces suffer from a narrow bandwidth, which restrict their further applications, especially in planar antenna design. To overcome this drawback, several methods have been proposed such as using stacked phase shifting elements or aperture patches coupled to true-time delay lines. Besides, there is another way to obtain a broadband working width by using a single-layer broadband planar antenna using ultrathin high-efficiency focusing metasurfaces.

In this chapter, a single-layer broadband focusing metasurface has been proposed to enhance the gain of the antenna. Theoretically, the spherical wave emitted by a point source at the focal point can be transformed to a plane wave. Therefore, a Vivaldi antenna has been fixed at the focal point of focusing metasurface, obtaining wideband planar antennas. In this case, the directivity and gain of the point source have been improved remarkably.

1.1 Element design

Figure 1 shows the proposed element, which is used to design reflective metasurface. The element is composed of orthogonally I-shaped structures and a metal-grounded plane spaced by a dielectric isolator with a permittivity of

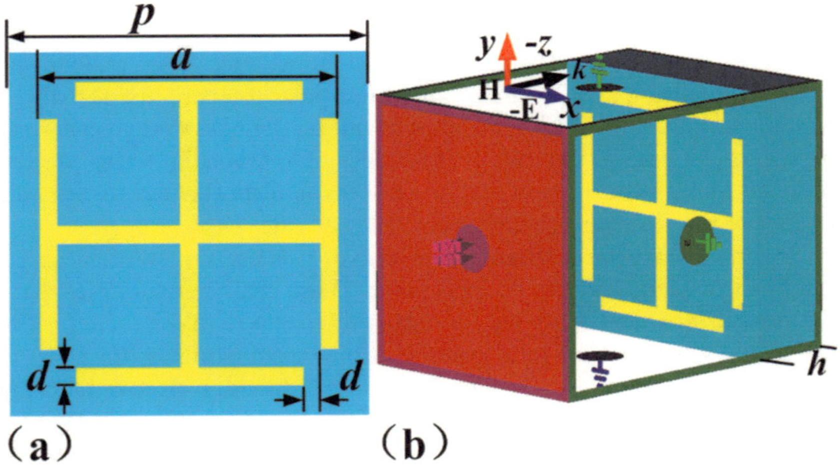

Figure 1.
Structure of the element and the simulated setup: (a) top view; (b) perspective view. The parameters are listed as d = 0.3 mm, p = 6 mm, h = 2 mm, and a = 1.8–5 mm.

ε_r = 2.65, loss tangent of 0.001, and thickness of 2 mm. For characterization, the element is simulated in commercial software CST Microwave Studio, unit cell boundary conditions are employed along x and y directions. Moreover, the element is illuminated by x-polarized plane wave along $-z$ direction.

Figure 2 shows the reflective phase of the reflection wave versus the length of parameter a. To demonstrate the ability of tuning the reflective phase at broadband, the reflective phases versus parameter a varying from 1.8 to 5 mm have been plotted from 15 to 22 GHz in **Figure 2**, respectively. Obviously, the phase tuning range is covering 2π at all frequencies (15–22 GHz). In additions, the eight phase shift curves nearly parallel to each other, leading to a large bandwidth for focusing EM wavefronts.

1.2 Broadband and focusing metasurface design

Generally, the reflected wave will always deflect to the phase delay direction according to the general reflection law as depicted in Eq. (1):

$$n_r \sin(\theta_r) - n_i \sin(\theta_i) = \frac{\lambda}{2\pi}\frac{d\Phi}{dx} \tag{1}$$

where Φ is the phase discontinuity at the local position on the metasurface, θ_r/θ_i is the reflected/incident angle, n_r/n_i is the reflective index of the reflected/incident medium, and λ is the wavelength. To clearly demonstrate the general reflection law, $d\Phi/dx$ can be denoted as $2\pi/np$, where n is the number of elements arranged in order along x direction and p is the periodicity of element. In such case, the element is illuminated normally, so the θ_i can be denoted by 0°. At the same time, the n_i = 1 can be realized because the element is placed in free space. Consequently, the reflected angle θ_r can be depicted in Eq. (2):

$$\theta_r = \sin^{-1}\left(\frac{\lambda}{2\pi} \times \frac{2\pi}{np}\right) \tag{2}$$

Then designing a focusing phase distribution on the metasurface by using the proposed element is the key procedure. Based on Fermat's principle, the EM

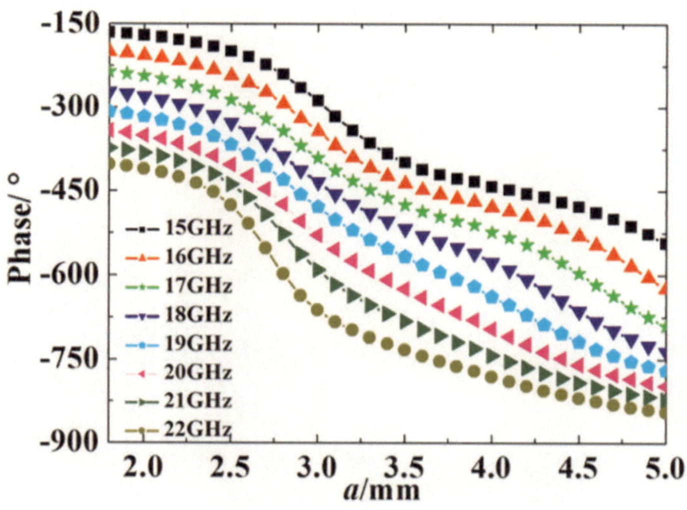

Figure 2.
Reflected phase shift with a (1.8–5 mm) from 15 to 22 GHz.

wavefront can be modified by changing the phase distribution on the metasurface. In order to focus the incident plane wave to a quasi-spherical wave, the phase $\Phi(m,n)$ imposed at location (m,n) should satisfy Eq. (3):

$$\Phi(m,n) = \frac{2\pi}{\lambda}\left(\sqrt{(mp)^2 + (np)^2 + L^2} - L\right) + \Phi_0 \tag{3}$$

where L is the focal length, Φ_0 is the phase of origin point (0,0), and p is the periodicity of the element. **Figure 3a** depicts the conversion of an incident plane wave to a quasi-sphere wave, and **Figure 3b** depicts the conversion of a quasi-sphere wave to plane wave using the focusing metasurfaces.

Based on the procedure, a focusing metasurface with a size of 90×90 mm^2, composed of 15×15 elements, is proposed and simulated. By theatrically calculating, a hyperbolic phase distribution is assigned on the metasurface. As shown in **Figure 4a** shows, the phase response along x direction is a hyperbolic phase distribution, and **Figure 4b** is a plot of the whole phase distribution on the metasurface.

In order to have an intuitionistic view of the focusing metasurface, the proposed metasurface, as shown in **Figure 5a**, is simulated in the commercial software CST.

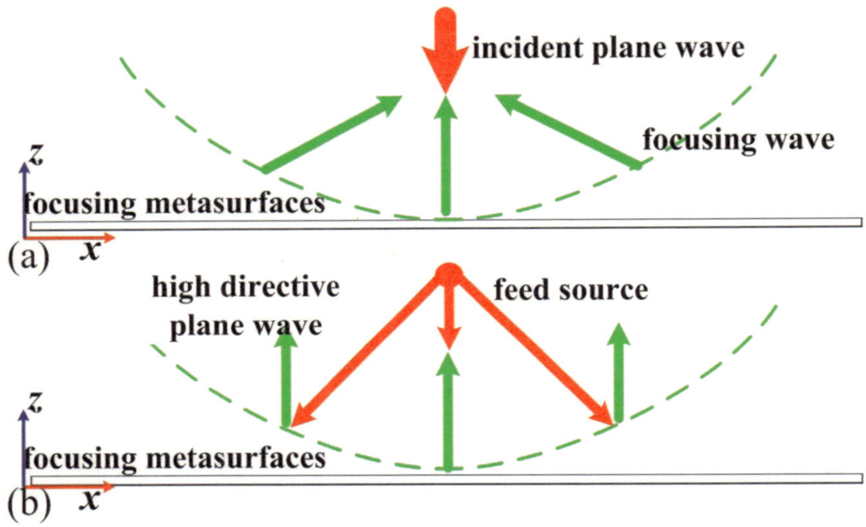

Figure 3.
(a) Schematic used to describe focusing effect and (b) schematic used to describe operating mechanism of planar antenna.

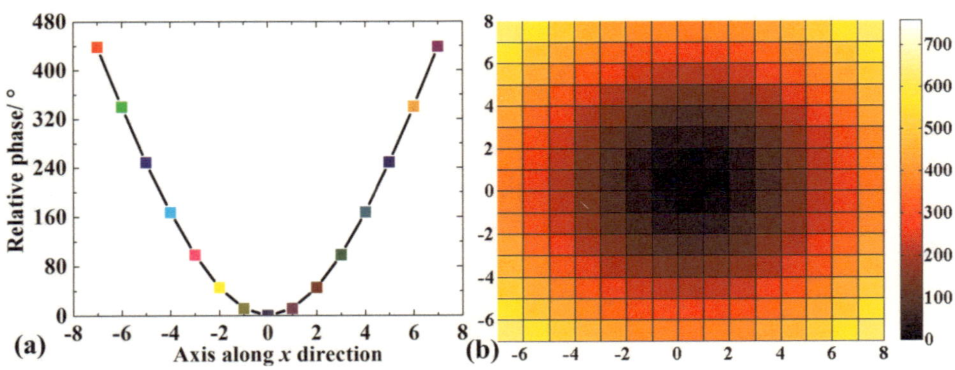

Figure 4.
(a) Phase response on the cut line along x direction and (b) relative reflection phase distribution in xoy plane.

The metasurface is illuminated by a plane wave with a polarization propagating along $-z$ direction. To verify the focusing effect, the electrical field at both *yoz* and *xoz* planes in the center frequency 18 GHz is plotted in **Figure 5b** and **c**. It is obvious that the incident plane wave is transformed into quasi-sphere wave in the orthogonal planes. Furthermore, to verify the position of the focus point, a curve is put along z direction, and power field is evaluated on the curve. As **Figure 5d** shown, the energy is focused at both *xoz* and *yoz* planes. The red spot is the focus point. Normalized power field versus the distance to the metasurface is described in **Figure 5d**. Observing the normalized power field, it is drawn that the focal point is at L = 33 mm, which agrees well with the theoretical calculation.

1.3 Broadband and high-gain planar antenna design

According to the above analysis, a spherical wave, emitted by a source located at focal point of the focusing metasurface, can be transformed into a plane wave theoretically. Therefore, a high-gain planar antenna can be realized by putting a feed antenna at the focal point of the focusing metasurface. The well-designed feed antenna is a Vivaldi antenna to offer a wide operating bandwidth. **Figure 6a** depicts

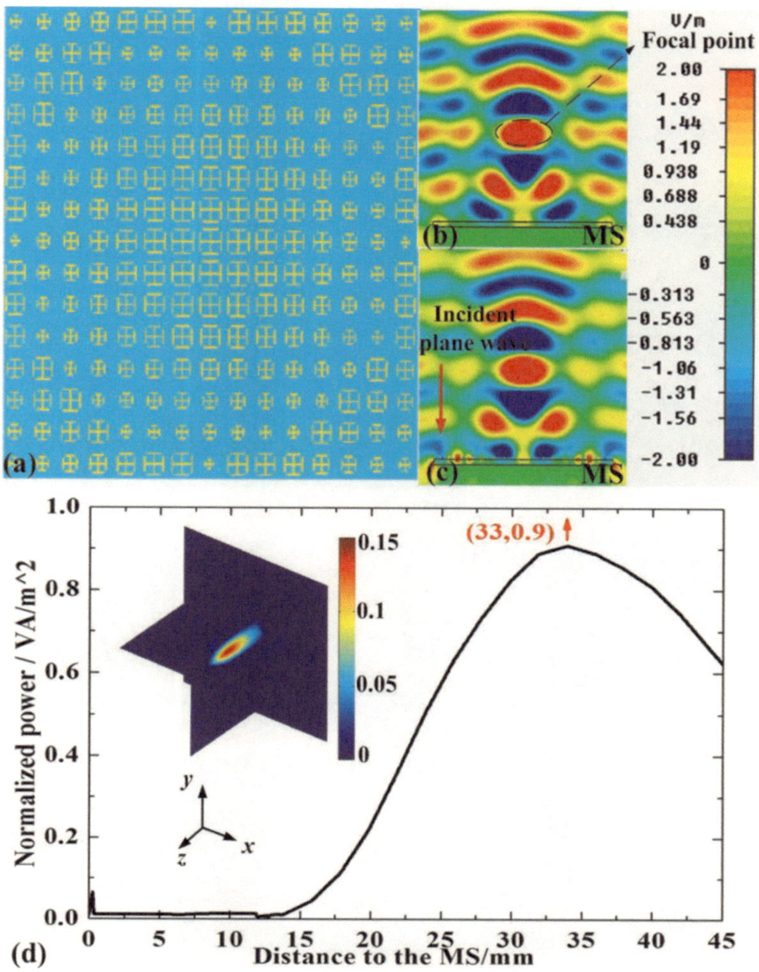

Figure 5.
(a) The simulated focusing MS; (b) simulated reflected electric field distribution in yoz plane at 18 GHz; (c) simulated reflected electric field distribution in xoz plane at 18 GHz; and (d) power distribution of focusing wave at 18 GHz and distance to the MS.

the geometry parameters of Vivaldi antenna, and **Figure 6b** plots the reflection coefficient. It is clear to find that the designed antenna can operate from 15 to 22 GHz with the S_{11} lower than −10 dB, indicating that the proposed feed antenna is a good feed source for the planar antenna.

To demonstrate the performance of the planar antenna, the simulated electrical field distributions with/without focusing metasurface at both *xoz* and *yoz* planes are

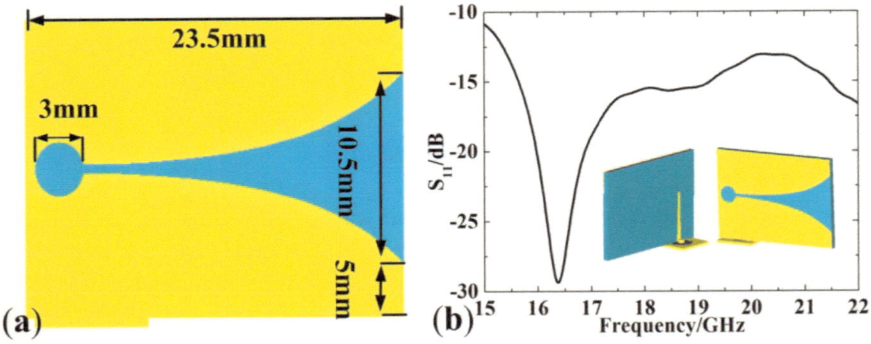

Figure 6.
(a) Parameters of Vivaldi antenna and (b) simulated S_{11} of Vivaldi antenna.

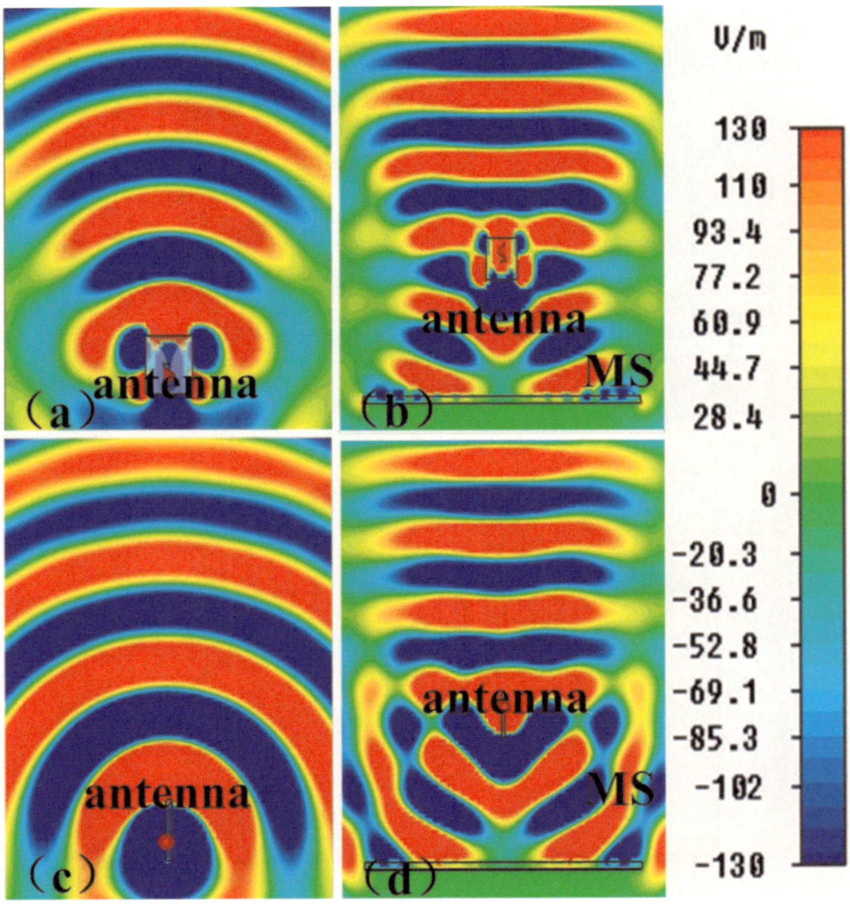

Figure 7.
Simulated electric field distribution at 15 GHz in (a and b) yoz plane and (c and d) xoz plane, respectively, for the Vivaldi antenna without (a and c) and (b and d) with the PGMS.

simulated. **Figures 7–9** plot the electrical field distributions at three representative frequencies (15, 18, and 22 GHz). As expected, the spherical wave emitted by feed antenna is transformed into nearly plane wave with the focusing metasurface in both *xoz* and *yoz* plane.

To clearly show the farfield performance of the planar antenna, the 3D radiation patterns at 15, 18, and 22 GHz are shown in **Figure 10**. The gain has been remarkably enhanced in a broad bandwidth comparing with the gain of the feed antenna. And pencil-shaped radiation pattern is achieved. Thus, the broadband and high-gain planar antenna is obtained. In order to verify the simulation, a sample composed of 15 × 15 elements is fabricated as **Figure 11a** shown. Besides, the designed Vivaldi antenna is fabricated and put at the focal point with a foam.

The simulated and measured radiation patterns in *xoz* plane and *yoz* plane at 18 GHz are plotted in **Figure 12**. It is obvious that the designed planar antenna has a remarkably enhanced gain compared to without focusing metasurface. The peak gain enhancement is about 15 dB relative to the bare Vivaldi antenna both in *xoz* plane and *yoz* plane. To demonstrate the broadband performance of the high-gain planar antenna, the simulated and measured gain versus frequency is described in **Figure 13**. It can be concluded that 3-dB gain bandwidth is from 15 to 22 GHz and the measured results conform to simulation well. Due to thin thickness, polarization

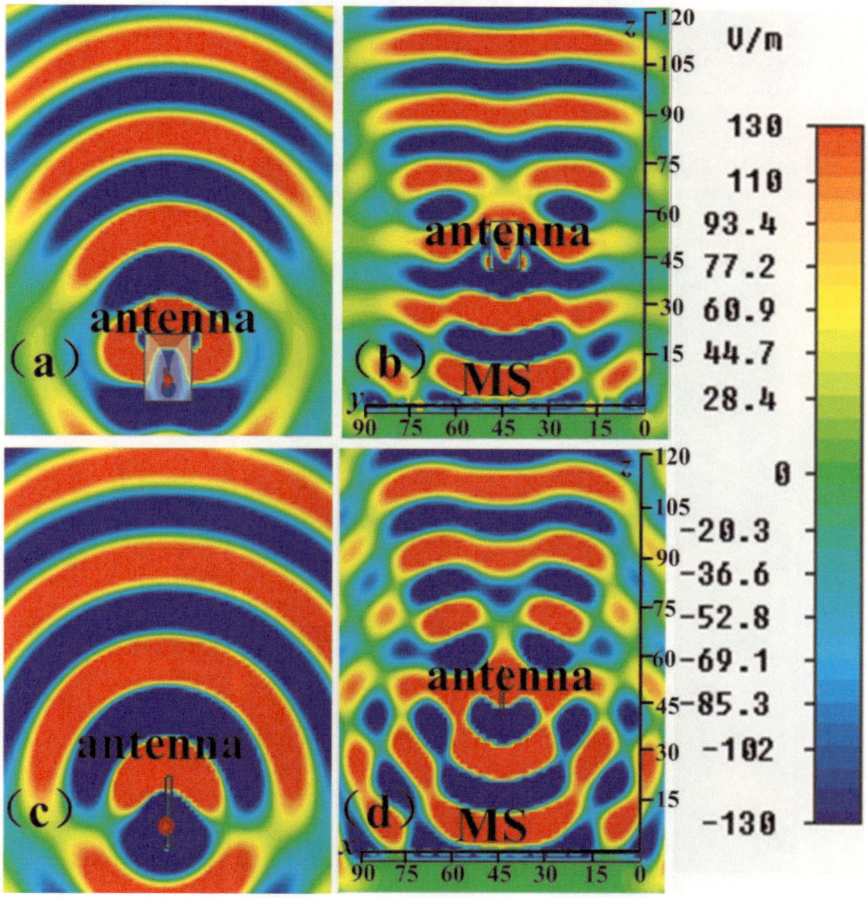

Figure 8.
Simulated electric field distribution at 18 GHz in (a and b) yoz plane and (c and d) xoz plane, respectively, for the Vivaldi antenna without (a and c) and (b and d) with the PGMS.

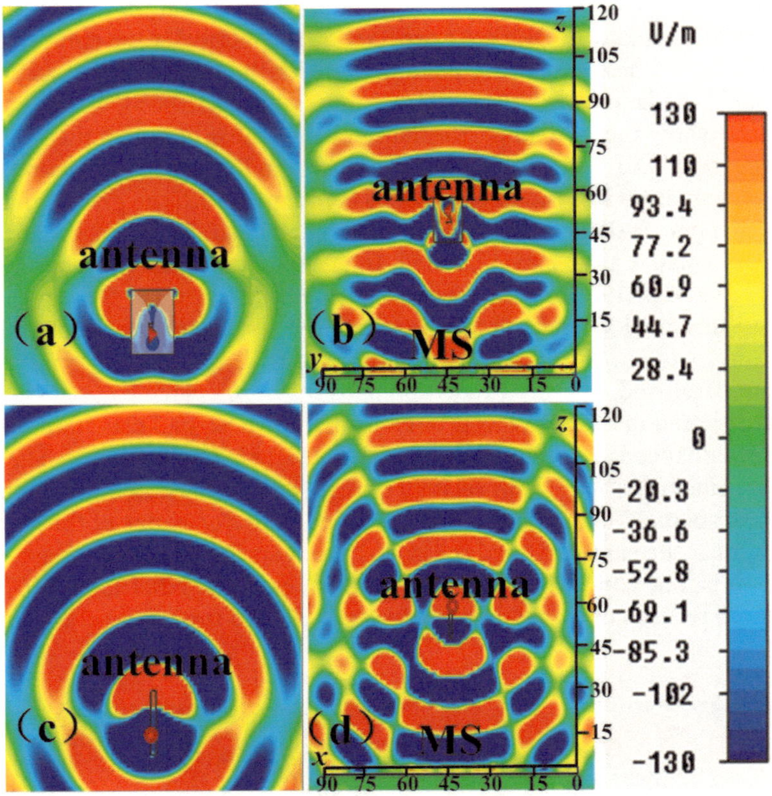

Figure 9.
Simulated electric field distribution at 22 GHz in (a and b) yoz plane and (c and d) xoz plane, respectively, for the Vivaldi antenna without (a and c) and (b and d) with the PGMS.

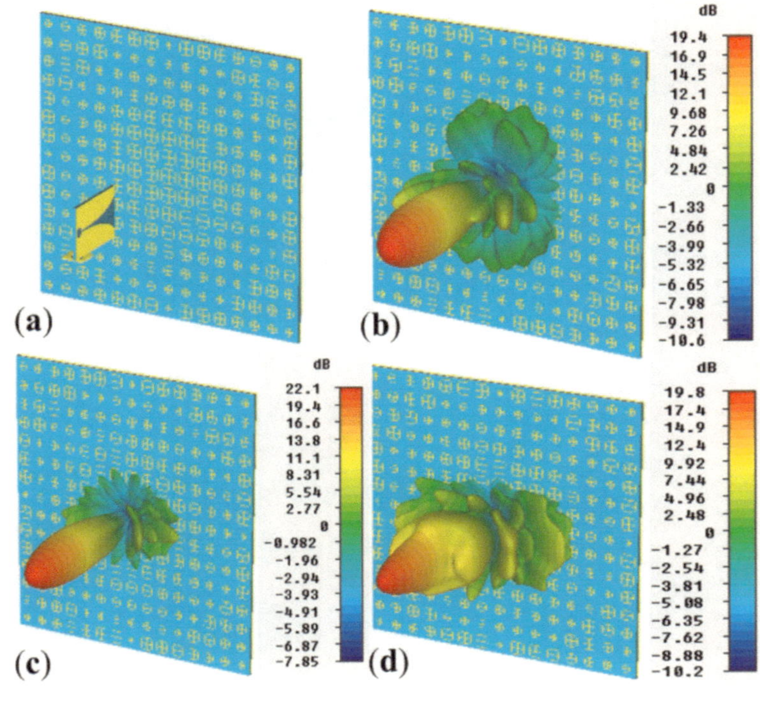

Figure 10.
(a) Simulated model for planar antenna and 3D radiation pattern for (b) 15 GHz; (c) 18 GHz; and (d) 22 GHz.

insensitivity, and broad bandwidth, the proposed broadband high-gain planar antenna opens up a new route for the applications of the metasurface in microwave band.

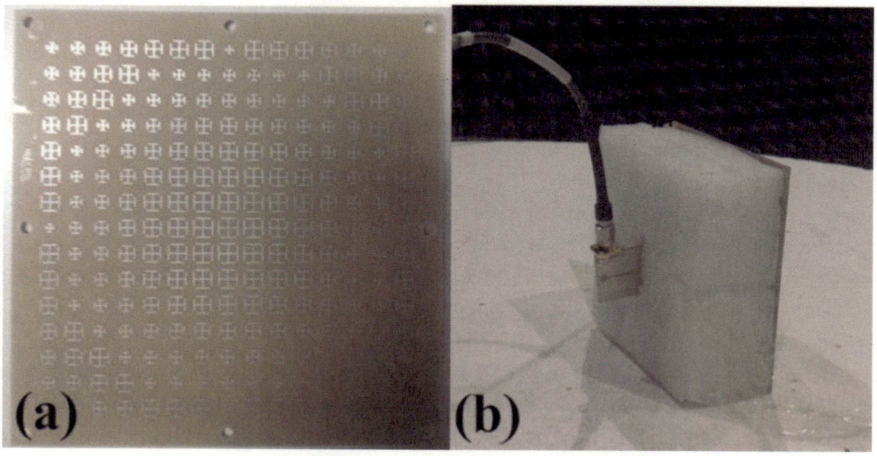

Figure 11.
The photographs of (a) metasurfaces top view and (b) planar high-gain antenna.

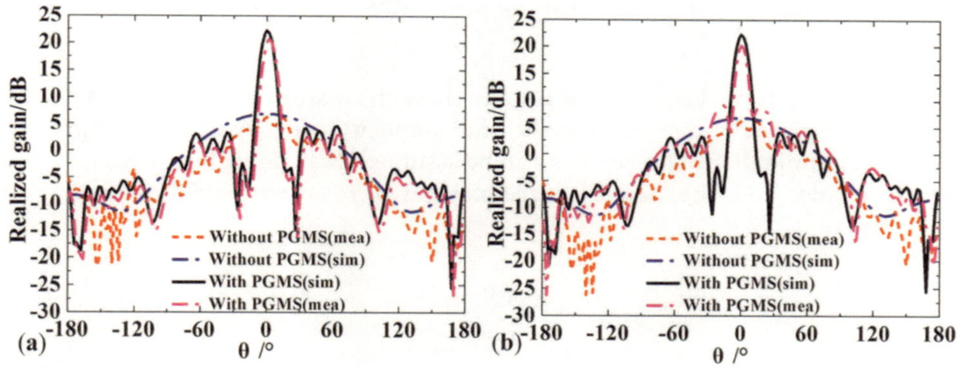

Figure 12.
Simulated and measured farfield radiation pattern at 18 GHz (a) xoz plane and (b) yoz plane.

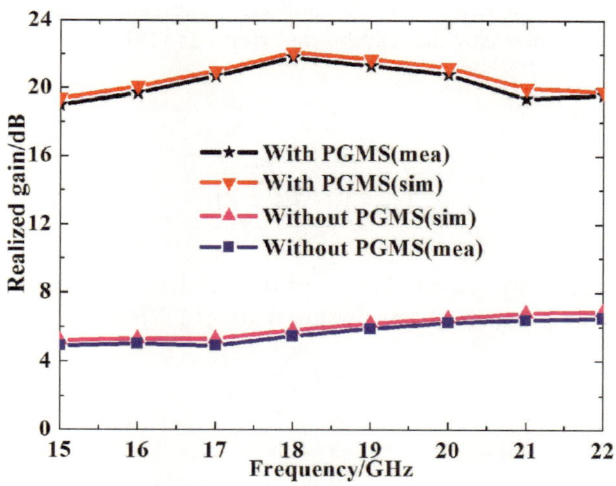

Figure 13.
Simulated and measured realized gain with/without PGMS from 15 to 22 GHz.

2. Highly efficient multifunctional metasurface for high-gain lens antenna application

With the development of the metasurface, it is a trend to design multifunctional devices to satisfy increasing requests of communication system in microwave region. Due to the ability of solving some key challenges like susceptibility to multipath, atmospheric absorptions, and reflections, circularly polarized antennas play an important role in wireless and satellite communication. At the same time, high-gain antenna plays an essential role in achieving long-distance wireless communication.

Owing to the function of linear-to-circular polarization conversion [13–17], metasurface opens up a novel route to realize the circular polarization. Therefore, it is interesting to design a circular antenna with high gain using metasurface. Generally, there are two categories of metasurface—transmitting type and reflecting type—according to the format of the metasurface. Compared with reflecting type, transmitting type allows reducing feed blockage effect when designing high-gain antenna. Therefore, it is more suitable for high-gain antenna. And it will be more novel to design a circularly polarized high-gain antenna engineered to realize linear-to-circular polarization conversion and EM waves focusing by transmitting metasurface.

2.1 Theoretical analysis of transmitted linear-to-circular polarization conversion

Assuming that the EM wave propagates through an arbitrary transmitted metasurface placed on xoz plane (z = 0), for simplicity, we only consider normal incidence illumination. Therefore, it can be assumed that an incident wave propagates along $-z$ direction. Then the incident wave's electric field can be described as

$$\vec{E_i} = (\hat{x}E_x + \hat{y}E_y) \tag{4}$$

$$\begin{pmatrix} E_x \\ E_y \end{pmatrix} = \begin{pmatrix} \cos\theta \\ \sin\theta \end{pmatrix} e^{-jkz} \tag{5}$$

where k is the wave number, θ is the angle along x axis, and complexes E_x and E_y represent the x-polarized and y-polarized states, respectively. The transmitted electric field through the metasurface can be described as [18]

$$\vec{E_t} = \left(\hat{x}E'_x + \hat{y}E'_y\right) \tag{6}$$

$$\begin{pmatrix} E'_x \\ E'_y \end{pmatrix} = T\begin{pmatrix} E_x \\ E_y \end{pmatrix} \tag{7}$$

Furthermore, the complex amplitudes of the incident and transmitted fields can be connected by the T matrix (transmission matrix) [19]:

$$T = \begin{pmatrix} T_{xx}T_{xy} \\ T_{yx}T_{yy} \end{pmatrix} \tag{8}$$

where T_{ij} represents the transmission coefficient of j-polarized incident wave and i-polarized transmission wave. Thus, the incident wave of linear polarization can be transformed to transmitted wave of a right-handed circular polarized (RHCP) or a left-handed circular polarized (LHCP).

As depicted in **Figure 14**, there are two typical cases of realizing linear-to-circular polarized conversion.

Take the case of θ = 45° as an example; the metasurface is illuminated with a linear polarization incident wave with the incident electric field E, titled 45° relative to $+x$ direction. Thus the vertical and horizontal components of E are set as

$$\begin{pmatrix} E_x \\ E_y \end{pmatrix} = \frac{1}{\sqrt{2}} \begin{pmatrix} 1 \\ 1 \end{pmatrix} e^{-jkz} \tag{9}$$

Therefore, the vertical and horizontal components of the transmitted electric field E can be denoted as

$$\begin{pmatrix} E'_x \\ E'_y \end{pmatrix} = T \frac{1}{\sqrt{2}} \begin{pmatrix} 1 \\ 1 \end{pmatrix} e^{-jkz} \tag{10}$$

To get the circular polarization, the amplitudes of E for vertical and horizontal components are ideally equal, and the phase of E for vertical and horizontal components experiences the distinct 90° phase shift, namely, $\text{Mag}(E'_x) = \text{Mag}\left(E'_y\right)$ and $\text{Arg}(E'_x) - \text{Arg}\left(E'_y\right) = -90°$. Therefore, $\text{Mag}(T_{xx}) = \text{Mag}(T_{yy}) = 1$, $\text{Mag}(T_{xy}) = \text{Mag}(T_{yx}) = 0$, and $\text{Arg}(T_{xx}) - \text{Arg}(T_{yy}) = -90°$ should be satisfied. At the same time, $\text{Arg}(T_{ij})$ represents the phase of j-polarized incident wave into i-polarized transmission wave. And $\text{Mag}(T_{ij})$ represents the amplitude of j-polarized incident wave into i-polarized transmission wave. In this case, the T matrix is

$$T = \begin{pmatrix} -j & 0 \\ 0 & 1 \end{pmatrix} e^{-j\varphi} \tag{11}$$

where φ is the phase obtained from the metasurface. Thereinto, the transmitted electric fields can be calculated as

$$\begin{pmatrix} E'_x \\ E'_y \end{pmatrix} = T \frac{1}{\sqrt{2}} \begin{pmatrix} 1 \\ 1 \end{pmatrix} e^{-jkz} e^{-j\varphi} = \frac{1}{\sqrt{2}} \begin{pmatrix} -j \\ 1 \end{pmatrix} e^{-jkz} e^{-j\varphi} \tag{12}$$

$$\vec{E_t} = \left(\hat{x}E'_x + \hat{y}E'_y\right) = \frac{1}{\sqrt{2}}(-j\hat{x} + \hat{y}) e^{-jkz} e^{-j\varphi} \tag{13}$$

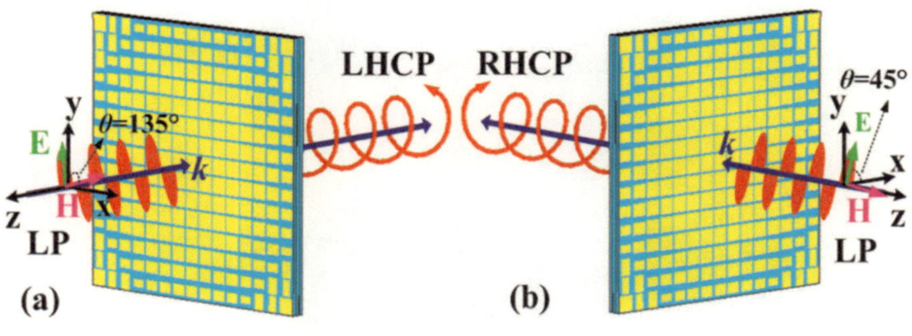

Figure 14.
The proposed two schematic models of LTC polarized for different incident electric fields at (a) θ = 135°; (b) θ = 45° relative to +x direction (θ is the angle between E and +x axis).

From Eq. (13), it can be concluded that a RHCP wave is obtained in this case as shown in **Figure 14b**.

In the other case of θ = 135° as shown in **Figure 14a**, the transmitted electric field can be calculated as

$$\begin{pmatrix} E'_x \\ E'_y \end{pmatrix} = T\frac{1}{\sqrt{2}}\begin{pmatrix} -1 \\ 1 \end{pmatrix} e^{-jkz}e^{-j\varphi} = \frac{1}{\sqrt{2}}\begin{pmatrix} j \\ 1 \end{pmatrix} e^{-jkz}e^{-j\varphi} \quad (14)$$

$$\vec{E_t} = \left(\hat{x}E'_x + \hat{y}E'_y\right) = \frac{1}{\sqrt{2}}(j\hat{x} + \hat{y})e^{-jkz}e^{-j\varphi} \quad (15)$$

Therefore, a LHCP wave will be obtained.

2.2 Design of the unit cell

Based on above theoretical analysis, it is necessary to design a unit cell with the ability of controlling x/y-polarized waves independently. The proposed unit cell, demonstrated in **Figure 15**, is composed of four metallic layers and three intermediate dielectric layers. Each metallic layer contains a same rectangular metal, which is employed to control the transmission phases and amplitudes of x/y-polarized EM waves. The dielectric layer has a substrate with the permittivity of ε_r = 4.3, loss tangent of 0.001, and thickness of h = 1 mm. The detailed parameters are denoted by h = 1 mm, p = 4.1 mm (periodicity of the unit cell), and a_y/a_x (1–3.8 mm).

To verify the polarization-independent property for x/y-polarized EM waves, the unit cell is illuminated by a plane wave propagating along $-z$ direction as shown in **Figure 15b**. Open boundary conditions and unit cell boundary conditions are applied in the z direction and x/y direction, respectively.

To demonstrate the ability of independently manipulating different polarized waves, the 2D map of phase shifts and amplitudes versus a_x and a_y is depicted in **Figures 16** and **17**. The phases and amplitudes of T_{xx} is plotted in **Figure 16**. As a_x and a_y vary from 1 to 3.8 mm, the phase of T_{xx} shown as **Figure 16a** nearly keeps a constant, indicating that the parameter a_y has no influence on the phase of T_{xx}, while the parameter a_x has an obvious influence on the phase of T_{xx}. As shown in **Figure 16b**, varying the parameter a_y, the amplitude of T_{xx} keeps a constant, indicating that a_y has no influence on the amplitude of T_{xx}. Therefore, it is can be drawn that T_{xx} can be controlled by varying the parameter a_x independently.

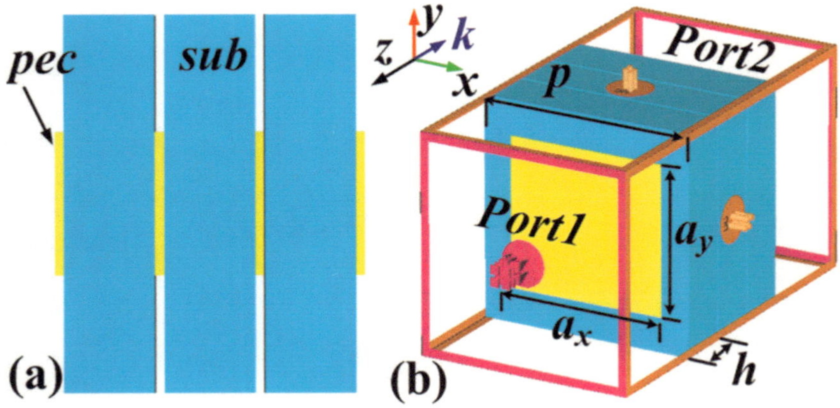

Figure 15.
Structure of the unit cell and simulated setup (a) top view and (b) perspective view.

DOI: http://dx.doi.org/10.5772/intechopen.88395

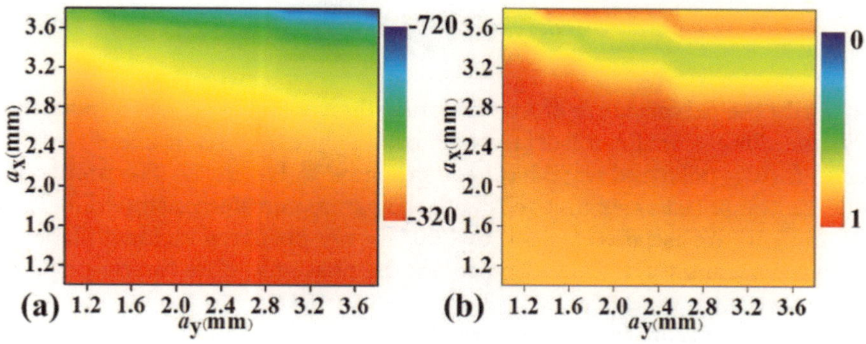

Figure 16.
Phases and amplitudes of T_{xx} of the unit cell.

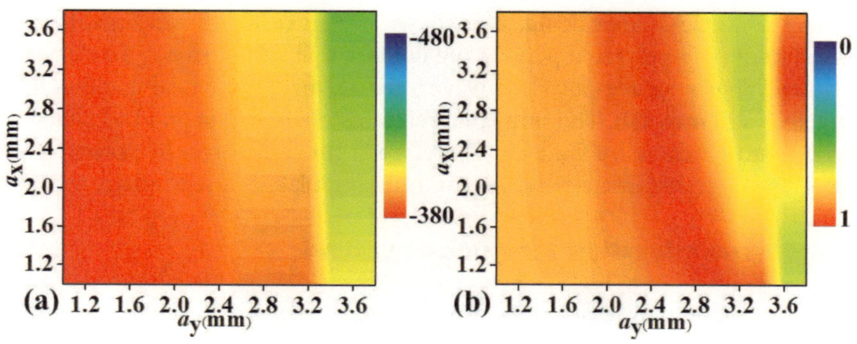

Figure 17.
Phases and amplitudes of T_{yy} of the unit cell.

Moreover, **Figure 17** shows the phase and amplitude of T_{yy} when the unit cell is illuminated by y-polarized wave when varying a_x and a_y from 1 to 3.8 mm. As a_x and a_y vary from 1 to 3.8 mm, the phase of T_{yy} shown as **Figure 17a** nearly keeps a constant, indicating that the parameter a_x has no influence on the phase of T_{yy}, while the parameter a_y has an obvious influence on the phase of T_{yy}. As shown in **Figure 17b**, varying the parameter a_x, the amplitude of T_{yy} keeps a constant, indicating that a_x has no influence on the amplitude of T_{yy}. Therefore, it is can be drawn that T_{yy} can be controlled by varying the parameter a_y independently.

Based on above analysis, it is concluded that the proposed unit cell has the ability of controlling x/y-polarized wave independently.

As we all know, the refracted wave will always deflect to the phase delay direction according to the general refraction law [3] as described in Eq. (16):

$$n_t \sin(\theta_t) - n_i \sin(\theta_i) = \frac{\lambda}{2\pi}\frac{d\varphi}{dx} \tag{16}$$

where φ is the phase discontinuity as the local position on the metasurface, $\theta_t(\theta_i)$ is the refracted (incident) angle of the EM waves, $n_t(n_i)$ is the refractive index of the transmissive (incident) medium, and λ is the wavelength. To simplify the analysis, $d\varphi/dx$ can be denoted as $2\pi/np$, where n is the number of the unit cell arranged along x direction and p is the periodicity of the unit cell. In the design, the unit cell is normally illuminated by EM wave in the free space; thus θ_i and n_t are denoted by 0° and 1, respectively. Thus the refraction angle θ_t can be depicted in Eq. (17):

$$\theta_t = \sin^{-1}\left(\frac{\lambda}{2\pi} \times \frac{2\pi}{np}\right) \tag{17}$$

To verify the performance of manipulating the x/y-polarized wave independently, one-dimensional metasurface with inverse linear phase gradient along x direction for x/y-polarized wave has been proposed in **Figure 17**. The proposed supercell is composed of six unit cells with the parameters demonstrated in **Table 1**. Phase gradients are assigned to +60° to −60° for x/y-polarized incident plane wave, respectively. Φ_x and Φ_y in **Table 1** represent the phase response for x/y-polarized incident waves.

4 × 16 supercells are fixed as **Figure 18a** has shown. The supercell is perpendicularly illuminated by plane wave along z direction in CST Microwave Studio. And open (add space) boundary condition is set along all directions.

Based on Eq. (17), the refracted angles for x/y-polarized incident wave can be calculated as 54.4°. And the refracted wave will deflect to $-x/+x$ direction, respectively. The 3D farfields are calculated as shown in **Figure 18a**. Beam 2 and beam 1 are calculated under x/y-polarized plane wave illumination, respectively. To clearly show the refracted angles for x/y-polarized waves, the normalized farfield patterns in polar are depicted in **Figure 18b**. The simulated refracted angles are in good accordance with theoretic ones calculated by Eq. (17). Therefore, a conclusion can be drawn that the phases of x/y-polarized incident waves can be manipulated by ax and ay, respectively.

2.3 Design of multifunctional transmission PGMS

Based on above unique property, it is easy to design a hyperbolic phase distribution on the multifunctional metasurface, which has the ability of polarization conversion and gain enhancement.

Index n	1	2	3	4	5	6
a_x (mm)	3.8	3.75	3.64	3.52	3.1	2.53
a_y (mm)	2.53	3.1	3.52	3.64	3.75	3.8
Φ_x (deg)	−704.3	−644.3	−584.3	−524.3	−464.3	−404.3
Φ_y (deg)	−404.3	−464.3	−524.3	−584.3	−644.3	−704.3

Table 1.
The sizes and absolute phase shifts of the six distributed unit cells.

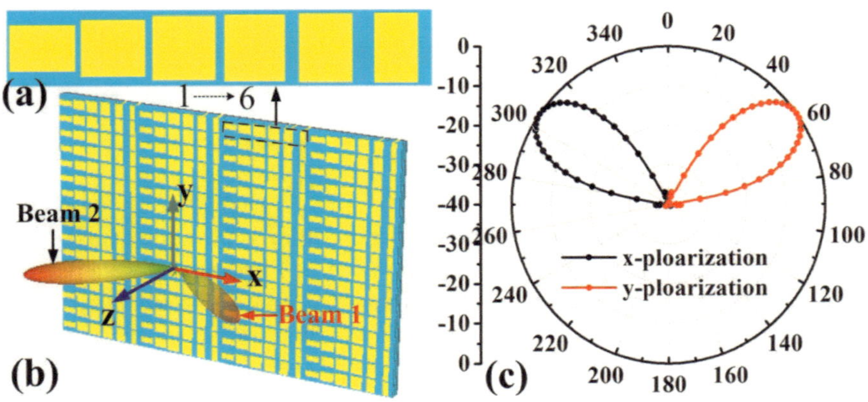

Figure 18.
(a) The supercells; (b) the designed one-dimensional phase gradient metasurface; and (c) the farfield results.

In order to efficiently convert the incident plane wave into a quasi-spherical wave, the phase $\varphi(m, n)$ at unit cell location (m, n) should be carefully optimized, yielding the following phase distribution:

$$\varphi(m,n) = \frac{2\pi}{\lambda}\left(\sqrt{(mp)^2 + (np)^2 + L^2} - L\right) + \varphi(0,0) \tag{18}$$

where p is periodicity of unit cell and L is the focal length and m and n are the unit cell positions along x and y directions, respectively. Moreover, $\varphi(0, 0)$ represents the phase at the origin unit cell (m = 0, n = 0). It is easy to convert the incident plane wave into quasi-spherical wave using the above metasurface. Moreover, we concentrate on designing a linear-to-circular polarization conversion metasurface according to the theory and unique feature described in the preceding section.

In order to realize the LTC conversion, the transmitted phase for x-polarized incident wave should have a phase difference of 90° with the y-polarized incident wave through each unit cell. In this chapter, we design a LTC conversion metasurface according to **Figure 14b**. Namely, $Arg(T_{xx})_{mn} - Arg(T_{yy})_{mn} \approx -90°$ should be satisfied, and a RHCP wave can be obtained. For x-polarized incident wave, the phase distribution can be denoted as

$$Arg(T_{xx})_{mn} - Arg(T_{xx})_{00} = \frac{2\pi}{\lambda}\left(\sqrt{\left((mp)^2 + (np)^2 + L^2\right)} - L\right) \tag{19}$$

For y-polarized incident wave, the phase distribution can be denoted as

$$Arg(T_{yy})_{mn} - Arg(T_{yy})_{00} = \frac{2\pi}{\lambda}\left(\sqrt{\left((mp)^2 + (np)^2 + L^2\right)} - L\right) \tag{20}$$

To achieve $Arg(T_{xx})mn - Arg(T_{yy})mn \approx -90°$ simply, the phase difference $Arg(T_{xx})_{00} - Arg(T_{yy})_{00} \approx -90°$ should be satisfied for x/y-polarized incident waves at origin point $\varphi(0, 0)$.

Based on above analysis, the multifunctional transmission metasurface is composed of 15 × 15 unit cells. The working frequency and focal length are f = 15 GHz and L = 30 mm, respectively. A patch antenna, operating from 14.5 to 15.3 GHz, is placed on the focal point to be a feed source. By insertion of p = 4.1 mm, λ = 20 mm,

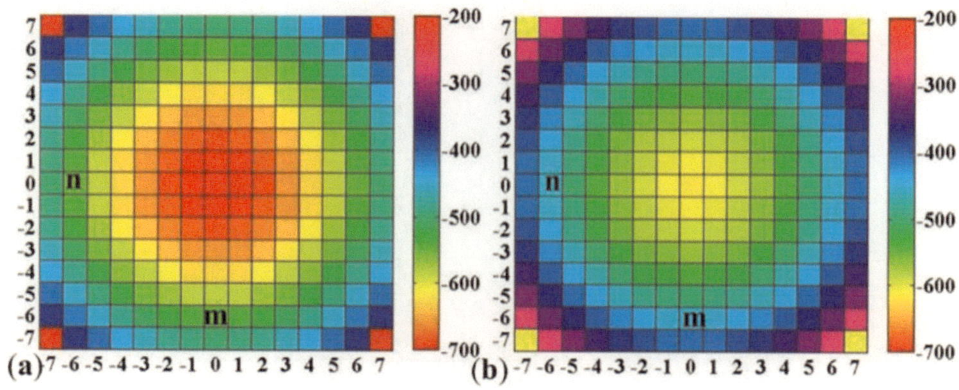

Figure 19.
Absolute phase distributions for (a) x polarization and (b) y polarization.

L = 30 mm, and $Arg(T_{xx})_{00} = -704.3°$ into Eqs. (19) and (20), the phase distribution for the x/y polarization waves is calculated as **Figure 19**. Finally, the proposed multifunctional transmission metasurface is depicted in **Figure 20a**.

To verify the proposed multifunctional transmission metasurface's function of focusing EM wave at L = 30 mm, the metasurface is simulated in CST. As shown in **Figure 20a**, the metasurface is normally illuminated by a plane wave with θ = 45° along the x axis, and a curve is put along z axis to evaluate power field on curve to calculate the focal point. Besides, a power flow monitor is set at 15 GHz. It is obvious that the plane wave is focused to a pink focal spot and the maximum power field is at L = 30 mm.

2.4 High-gain lens antenna design

A designed patch antenna, operating at 15 GHz, is put at the focal point of the multifunctional metasurface. As **Figure 20** has shown, the polarization of EM wave emitted by the feed source antenna has an angle of θ = 45° along the x axis. To clearly show the function of proposed lens antenna, simulated 3D radiation pattern at 15 GHz and measured S_{11} of the lens antenna are plotted in **Figure 21**. It is

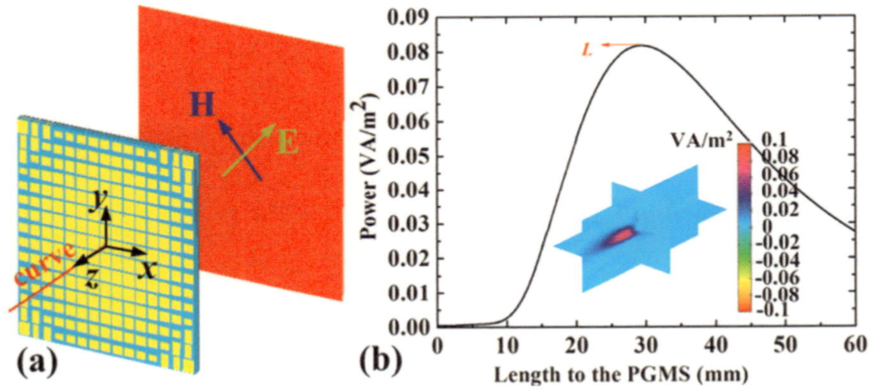

Figure 20.
(a) The multifunctional transmission PGMS and simulated conditions and (b) power field distribution in xoz and yoz planes and power field distribution along z axis.

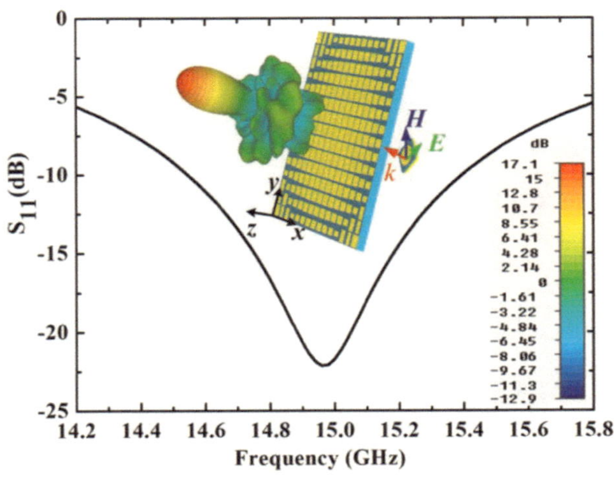

Figure 21.
The simulated 3D radiation pattern and measured S_{11} of lens antenna.

obvious that the pencil-beam 3D radiation pattern is obtained and realized gain has been enhanced remarkably.

Furthermore, the near-field electric field distributions in *xoz* and *yoz* planes have been depicted in **Figure 22**. The incident wave has been converted into near-plane wave through the proposed metasurface. The simulated result is in line with our expectation.

Lastly, a sample is fabricated and measured in a microwave anechoic chamber as shown in **Figure 23**. And simulated and measured 2D radiation patterns of lens antenna are plotted in **Figure 24**. As shown in **Figure 24**, the co-polarization and cross-polarization of simulation and measurement are plotted in two orthogonal planes. And the simulated results are in good accordance with the measured ones. Compared with patch antenna, the radiation patterns of proposed multifunctional antenna are more directional. The measured peak gain of patch antenna and proposed multifunctional antenna is about 5.9 and 16.9 dB at 15 GHz, respectively. And the realized gain has been enhanced with 11 dB at 15 GHz, and calculated aperture efficiency is about 41.2%. Comparing co-polarization with cross-polarization at $\theta = 0°$, we find that the isolation is better than 17 dB.

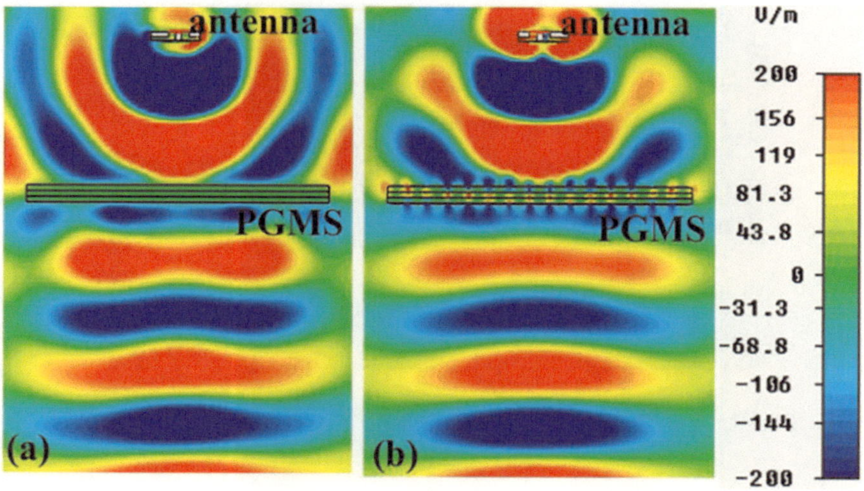

Figure 22.
Electric field distributions in (a) xoz and (b) yoz plane at 15 GHz.

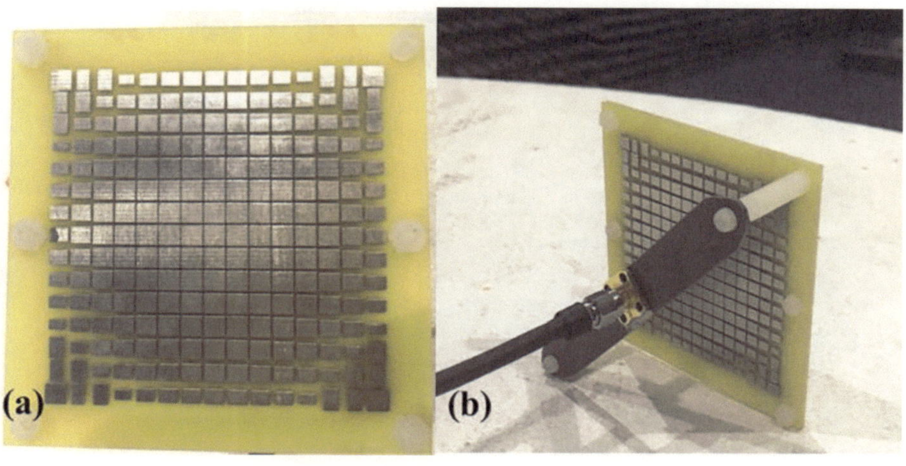

Figure 23.
The photographs of (a) PGMS top view and (b) lens antenna.

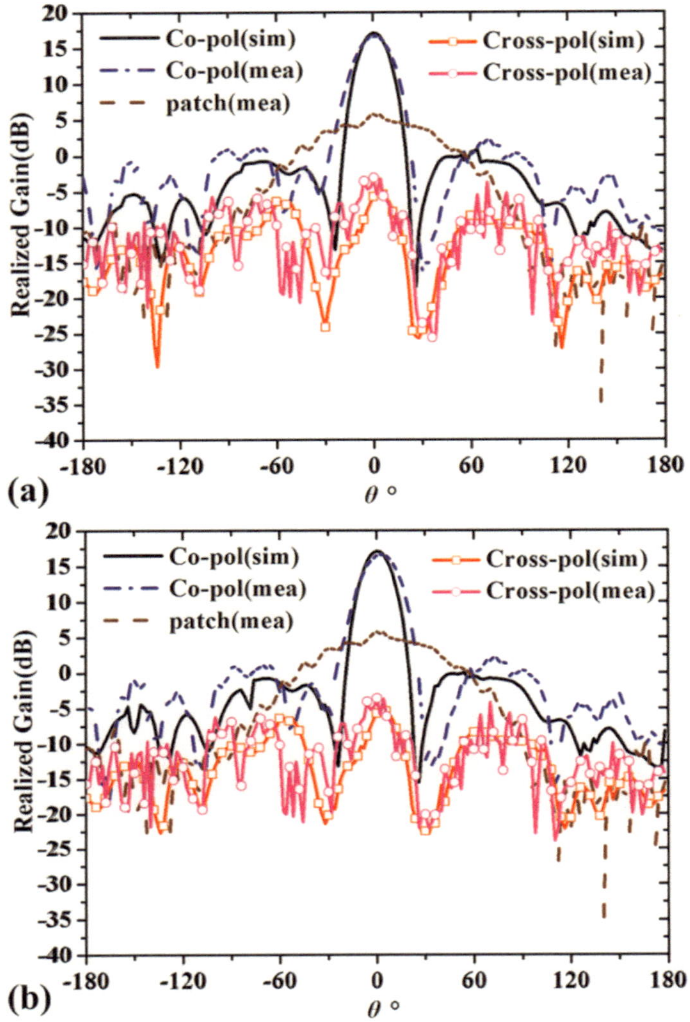

Figure 24.
2D radiation patterns at 15 GHz. (a) xoy plane and (b) yoz plane.

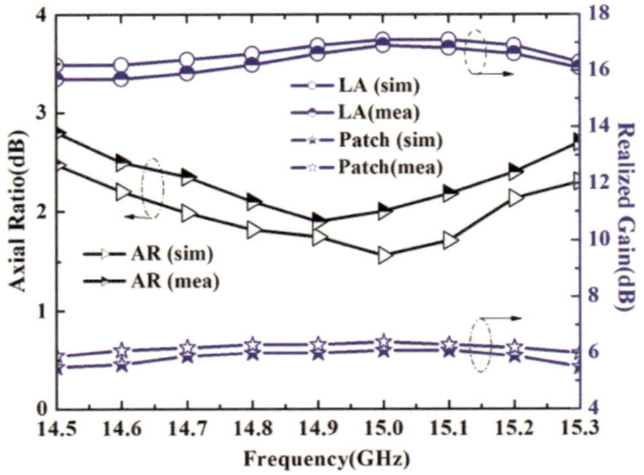

Figure 25.
Simulated and measured realized gain for patch antenna and lens antenna and axial ratio.

Moreover, the simulated and measured axial ratios (AR)($\theta = 0°$) are plotted in **Figure 25**. The simulated and measured realized gain of lens antenna and patch antenna is described in **Figure 25**. All the simulated and measured results are in good accordance. It is obvious that the gain enhancement is about 11 dB at 15 GHz. Besides, the axial ratio bandwidth for AR < 3 dB ranges from 14.5 to 15.3 GHz with the fractional bandwidth 5.3%. And we find that the 1-dB gain bandwidth is 5.3% (14.5–15.3 GHz), which agrees well with the axial ratio bandwidth.

Conclusions

In this chapter, we have reviewed our recent efforts in utilizing metasurface to enhance the performance of the conventional antenna. For the reflected metasurface, we propose a new broadband, single-layered reflected focusing metasurface, and take it into application for high-gain planar antenna. The metasurface exhibits good focusing phenomenon from 15 to 22 GHz. Both simulation and measured results show that the peak gain of planar antenna has been averagely enhanced by 16 and −1 dB gain bandwidth which is from 15 to 22 GHz, while for the transmitted metasurface, a novel multifunctional metasurface combining linear-to-circular polarization conversion and EM waves focusing has been proposed and applied to designed a high-gain lens antenna. A RHCP lens antenna is simulated and measured. The measured results show that the lens antenna can convert LP waves into RHCP waves at 15 GHz. The 3-dB axial ratio bandwidth is 5.3%. Realized gain at 15 GHz is 16.9 dB, corresponding to aperture efficiency of 41.2%. These above metasurface antennas not only open up a new route for the applications of focusing metasurfaces in microwave band but also afford an alternative for high-performance antennas.

Acknowledgements

The authors would like to express their gratitude to anonymous reviewers for their helpful comments and China North Electronic Engineering Research Institute for the fabrication. This work was supported by the National Natural Science Foundation of China (Grant Nos. 61372034).

Author details

Haisheng Hou, Haipeng Li*, Guangming Wang*, Tong Cai, Xiangjun Gao and Wenlong Guo
Microwave Laboratory, Air Force Engineering University, Xi'an, China

*Address all correspondence to: s_lihaipeng@sina.cn and wgming01@sina.com

References

[1] Li HP, Wang GM, Gao XJ, Liang JG, Hou HS. Novel metasurface for dual-mode and dual flat high-gain antenna application. IEEE Transactions on Antennas and Propagation. 2018;**66**(7): 3706-3711

[2] Li HP, Wang GM, Cai T, Liang JG, Gao XJ. Phase- and amplitude-control metasurfaces for antenna main-lobe and sidelobe manipulations. IEEE Transactions on Antennas and Propagation. 2018;**66**(10):5121-5129

[3] Biswas SR, Gutierrez CE, Nemilentsau A, Lee IH, Oh SH, Avouris P, et al. Tunable graphene metasurface reflectarray for cloaking, illusion, and focusing. Physical Review Applied. 2018;**9**(3):034021

[4] Liang JJ, Huang GL, Zhao JN, Gao ZJ, Yuan T. Wideband phase-gradient metasurface antenna with focused beams. IEEE Access. 2019;7:20767-20772

[5] Katare KK, Chandravanshi S, Biswas A, Akhtar MJ. Realization of split beam antenna using transmission-type coding metasurface and planar lens. IEEE Transactions on Antennas and Propagation. 2019;**67**(4):2074-2084

[6] Yang WC, Gu LZ, Che WQ, Meng Q, Xue Q, Wan C. A novel steerable dual-beam metasurface antenna based on controllable feeding mechanism. IEEE Transactions on Antennas and Propagation. 2019;**67**(2): 784-793

[7] Yu NF, Genevet P, Kats MA, Aieta F, Tetienne J-P, Capasso F, et al. Light propagation with phase discontinuities: Generalized laws of reflection and refraction. Science. 2011;**334**:333-337

[8] Wong AMH, Eleftheriades GV. Perfect anomalous reflection with a bipartite Huygens' metasurface. Physical Review X. 2018;**8**(1):011036

[9] Xu WK, Zhang M, Ning JY, Wang W, Yang TZ. Anomalous refraction control of mode-converted elastic wave using compact notch-structured metasurface. Material Research Express. 2019;**6**(6): 065802

[10] Asadpor L, Sharifi G, Rezvani M. Design of a high-gain wideband antenna using double-layer metasurface. Microwave and Optical Technology Letters. 2019;**61**(4):1001-1010

[11] Ling YH, Huang LR, Hong W, Liu TJ, Jing L, Liu WB, et al. Polarization-switchable and wavelength-controllable multi-functional metasurface for focusing and surface-plasmon-polariton wave excitation. Optics Express. 2017;**25**(24): 29812-29821

[12] Meng YY, Ma H, Li YF, Feng MD, Wang JF, Li ZQ, et al. Spoof surface plasmon polaritons excitation and wavefront control by Pancharatnam-Berry phase manipulating metasurface. Journal of Physics D: Applied Physics. 2018;**51**(21):215302

[13] Lin BQ, Guo JX, Chu P, Huo WJ, Xing Z, Huang BG, et al. Multiple-band linear-polarization conversion and circular polarization in reflection mode using a symmetric anisotropic metasurface. Physical Review Applied. 2018;**9**:024038

[14] Jia YT, Liu Y, Zhang WB, Wang J, Wang YZ, Gong SX, et al. Ultra-wideband metasurface with linear-to-circular polarization conversion of an electromagnetic wave. Optical Materials Express. 2018;**8**(3):597-604

[15] Zheng Q, Guo CJ, Ding J. Wideband metasurface-based reflective polarization converter for linear-to-linear and linear-to-circular polarization conversion. IEEE Antennas and

Wireless Propagation Letters. 2018; **17**(8):1459-1463

[16] Khan MI, Khalid Z, Tahir FA. Linear and circular-polarization conversion in X-band using anisotropic metasurface. Scientific Reports. 2019;**9**:4552

[17] Ratni B, de Lustrac A, Plau GP, Burokur SN. Electronic control of linear-to-circular polarization conversion using a reconfigurable metasurface. Applied Physics Letters. 2017;**111**(21):214101

[18] Wang W, Guo ZY, Ran LL, Sun YX, Shen F, Li Y, et al. Polarization-independent characteristics of the metasurfaces with the symmetrical axis's orientation angle of 45° or 135°. Journal of Optics. 2016;**18**:035007

[19] Menzel C, Rockstuhl C, Lederer F. Advanced Jones calculus for the classification of periodic metamaterials. Physical Review A. 2010;**82**:053811

Chapter 4

Fractal Antennas for Wireless Communications

Amer T. Abed, Mahmood J. Abu-AlShaer and Aqeel M. Jawad

Abstract

When the length of the antenna is less than a quarter of the wavelength of the operating frequency, good radiation properties are difficult to obtain. However, size limitations can be overcome in this case using a fractal geometry antenna. The shape is repeated in a limited size such that the total length of the antenna is increased to match, for example, half of the wavelength of the corresponding desired frequency. Many fractal geometries, e.g., the tree, Koch, Minkowski, and Hilbert fractals, are available. This chapter describes the details of designing, simulations, and experimental measurements of fractal antennas. Based on dimensional geometry in terms of desired frequency bands, the characteristics of each iteration are studied carefully to improve the process of designing the antennas. In depth, the surface current distribution is investigated and analyzed to enhance the circular polarization radiation and axial ratio bandwidth (ARBW). Both, simulation and experimental, results are discussed and compared. Two types of fractal antennas are proposed. The first proposed fractal antenna has a new structure configured via a five-stage process. The second proposed fractal antenna has a low profile, wherein the configuration of the antenna was based on three iterations.

Keywords: fractal antenna, compact size, circular polarization

1. Introduction

Fractal means broken or irregularly fragmented and refers to a family of complex shapes that possess an inherent self-similarity in their geometrical structures. Good radiation properties are difficult to obtain when the length of the antenna is less than a quarter of the wavelength for the operating frequency. However, size limitations can be overcome in this case using a fractal geometry antenna. The shape is repeated in a limited size such that the total length of the antenna is increased to match, for example, half of the wavelength of the corresponding desired frequency. Many fractal geometries, such as tree, Koch, Minkowski, and Hilbert fractals, are used in designing antennas [1]. But through using a fractal geometry antenna, where the shape is repeated in a limited size, in a way that increases the total length of the antenna to match, for example, half of the wavelength of the corresponding desired frequency, the size limitations can be overcome in this case. Many fractal geometries such as the tree, Koch, Minkowski, and Hilbert fractals are used in designing this type of antenna.

There are many geometries used in designing fractal antennas such as: Fractal slot [2], Giuseppe Peano Fractal Geometries [3], Fractal loop [4], Fractal Cantor [5],

Minkowski Fractal [6], Koch Fractal [7], H-Fractal [8], Sierpinski gasket arrangement [9], Fern Fractal leaf [10], Mandelbrot Fractal antenna [11], Amer fractal slot [12], Sunflower Fractal [13], Flame Fractal [14] and Butterfly Structure [15].

2. Compact fractal antenna

One of the most widely used structures in fractal antennas is Sierpinski gasket [16]; it consists of equilateral triangles. There are two ways to build this structure, either by the decomposition method or by a multiple copy method. In this research, the second method is used in designing the fractal antenna; **Figure 1** represents this method.

At the first iteration, the structure made two copies of the same triangle in the 0th iteration, one of them located on its side while the other located above them. In the second iteration the same process is repeated, but with all first iteration structure. So the dimensions of the next iteration increased by factor 3 compared with the dimensions of the previous iteration. The above transformation of the triangle to generate any order of iterations can be represented by the mathematical formula [16]:

$$W\begin{bmatrix} x \\ y \end{bmatrix} = \begin{bmatrix} r\,cos\theta & -s\ cos\,\varnothing \\ r\,sin\theta & s\ sin\,\varnothing \end{bmatrix}\begin{bmatrix} x \\ y \end{bmatrix} + \begin{bmatrix} x_0 \\ y_0 \end{bmatrix} \tag{1}$$

where r and s are the scale factor, θ and $\varnothing$ are the rotation angles, and x_0 and y_0 are the amounts of translation. If the factors r, s are either reductions or magnifications, the transformation process is called self-affine, while, if $r = s$ and $\theta = \varnothing$, the transformation is called self-similar.

The structure of the proposed antenna is new; the initiator is a square patch as shown in **Figure 2a**; the two arms of the patch are equal and unity, i.e., $x = y = 1$. In 0th iteration, four symmetrical slots are cut in the square patch as shown in **Figure 2b**. The modified patch (**Figure 2c**) is configured by cutting increasable slips to change the dimensions of the arms at the corners in a way to configure asymmetrical corner dimensions:

$$W\begin{bmatrix} x \\ y \end{bmatrix} = \begin{bmatrix} r\,cos\theta & -s\ cos\,\varnothing \\ r\,sin\theta & s\ sin\,\varnothing \end{bmatrix}\begin{bmatrix} x \\ y \end{bmatrix} + \begin{bmatrix} x_0 \\ y_0 \end{bmatrix} \tag{2}$$

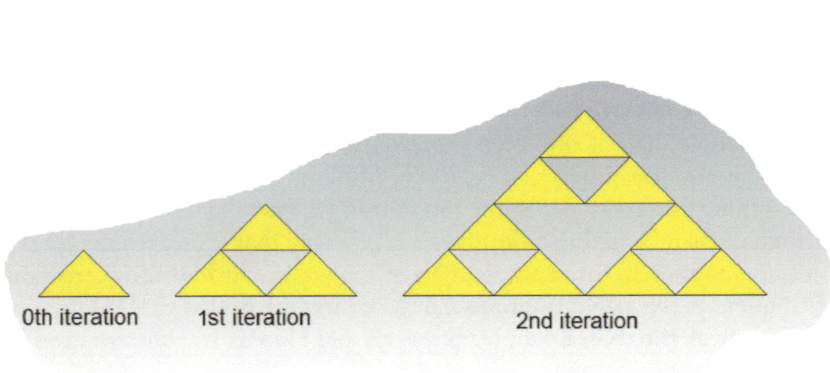

Figure 1.
Multiple copy approach of Sierpinski gasket arrangement.

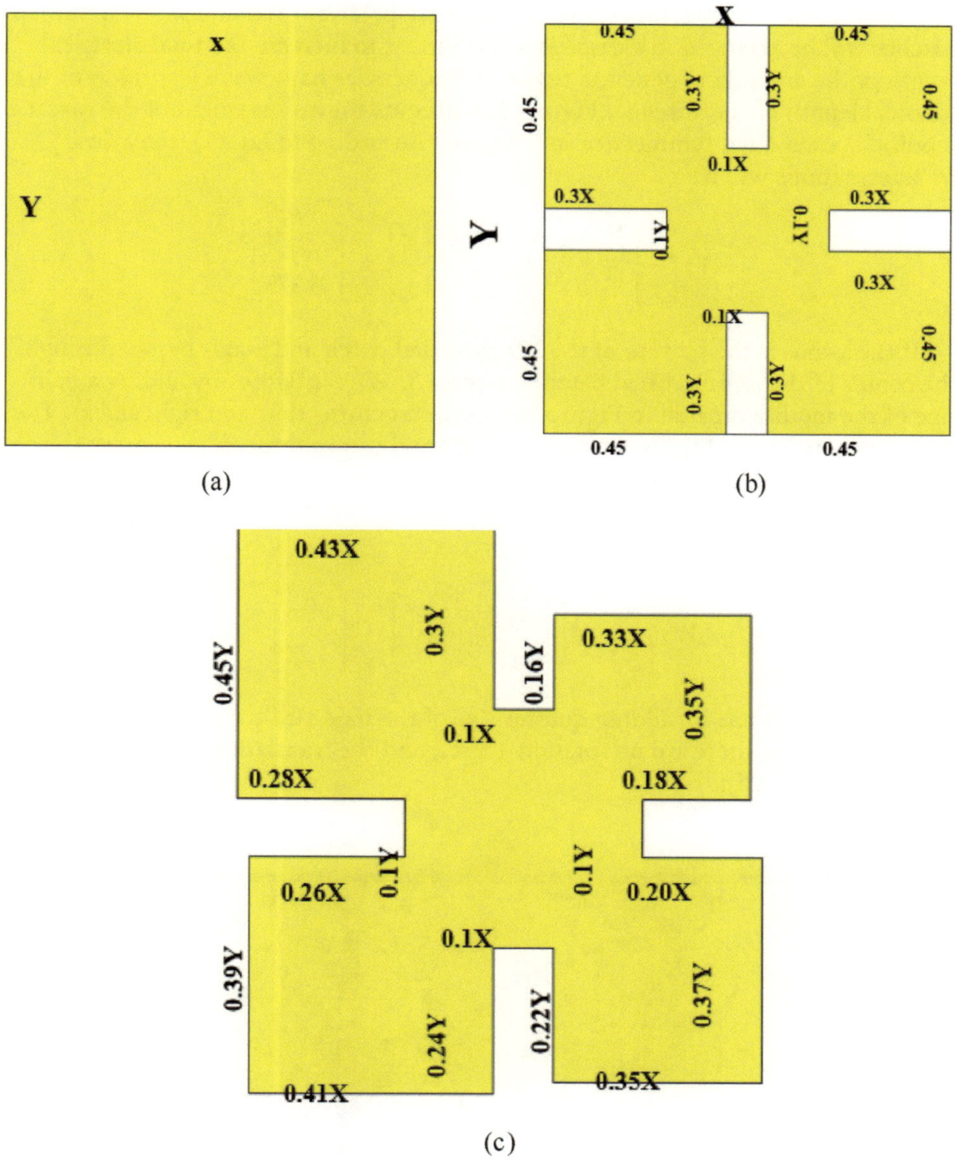

Figure 2.
Configuration of fractal antenna. (a) Initiator patch. (b) oth iteration. (c) Modified patch.

The upper arm in **Figure 2c** has total length $0.43x + 0.3y + 0.1x + 0.16y + 0.33x$ $=1.32x$, where $x = y$. The left arm's length is $0.45y + 0.28x + 0.1y + 0.26 +$ $0.39 = 1.48x$. The lower arm length is $0.41x + 0.24y + 0.22y + 0.1x + 0.35x = 1.32x$. The right arm length is $0.37y + 0.2x + 0.1y + 0.18x + 0.35y = 1.2x$. So, each arm in the square patch (**Figure 2a**) has unity length ($x = y$), while the lengths of the arms in the modified patch expand to $1.32x$, $1.4x$, and $1.2x$. This expedition gives two additional properties in designing the fractal antenna.

Firstly, different electrical lengths of the arms generate many resonant frequencies which can be integrated to have wide operating bands.

Secondly, different lengths of both sides of each corner, for example, the lengths of the two sides of the right-upper corner, are $0.35y$ and $0.33x$; this difference in lengths is useful to generate two orthogonal modes with phase shift 90° which are very important requirements to create circularly polarized radiation. The modified patches will be arranged in a cascade arrangement to increase the total electrical length of the antenna to generate resonant frequencies have wave lengths ≫ of the physical length of the antenna. **Figure 4** represents the arrangement of the cascade modified patch (two symmetrical structures). According to Eq. (1), the affine transformations will be:

$$W_2\begin{bmatrix} x \\ y \end{bmatrix} = \begin{bmatrix} 0.25 & 0.25 \\ 0 & 0 \end{bmatrix}\begin{bmatrix} x \\ y \end{bmatrix} + \begin{bmatrix} 0 \\ 6.375 \end{bmatrix} \tag{3}$$

If the x-axis is the bottom of the left modified patch and y-axis passes through the center of the left modified patch in **Figure 3**, W_1 configures by adding a half size of the modified patch in **Figure 3** to both structures (left and right sides). The dotted yellow line in **Figure 3** is on the y-axis and passes through all centers (black points) of the transformation structures (W_1, W_2 and W_3), so the scale factors are ($r = s = 0.5$), the rotation angles are $\theta = \varnothing = 0$, and the translation factors are ($x_0 = 0, y_0 = 4.25$):

$$W_1\begin{bmatrix} x \\ y \end{bmatrix} = \begin{bmatrix} 0.5 & 0.5 \\ 0 & 0 \end{bmatrix}\begin{bmatrix} x \\ y \end{bmatrix} + \begin{bmatrix} 0 \\ 4.25 \end{bmatrix} \tag{4}$$

While W_2 is done by adding quarter size of the modified patch, the scale factors are ($r = s = 0.25$), there are no rotation angles, and the translation factors are ($x_0 = 0, y_0 = 6.375$):

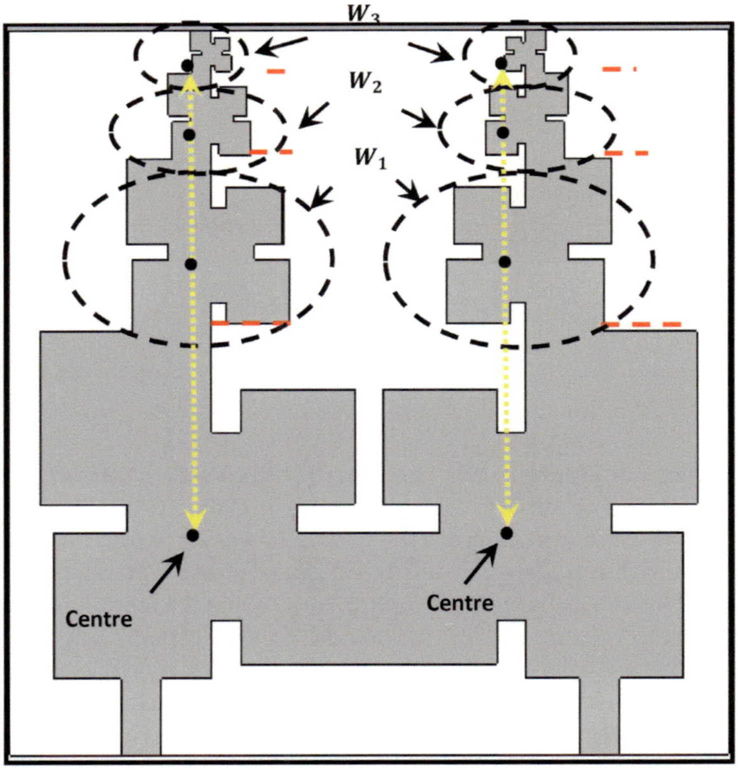

Figure 3.
Fractal dual-input antenna.

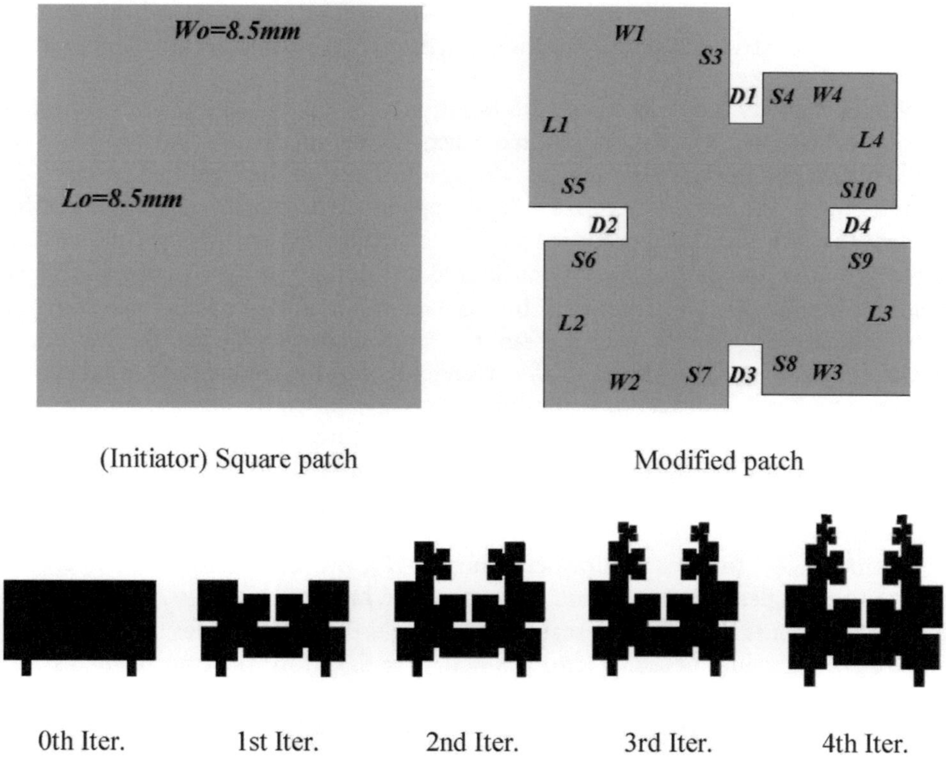

Figure 4.
Process of fractal antenna configuration.

$$W_2\begin{bmatrix} x \\ y \end{bmatrix} = \begin{bmatrix} 0.25 & 0.25 \\ 0 & 0 \end{bmatrix}\begin{bmatrix} x \\ y \end{bmatrix} + \begin{bmatrix} 0 \\ 6.375 \end{bmatrix} \tag{5}$$

And for W_3, the scale factors are ($r = s = 0.1250$) and the translation factors are ($x_0 = 0, y_0 = 7.4375$):

$$W_3\begin{bmatrix} x \\ y \end{bmatrix} = \begin{bmatrix} 0.125 & 0.125 \\ 0 & 0 \end{bmatrix}\begin{bmatrix} x \\ y \end{bmatrix} + \begin{bmatrix} 0 \\ 7.4375 \end{bmatrix} \tag{6}$$

The antenna had dual operating bands that meet the specifications of the Wi-Fi and WiMAX applications. The structure of the antenna was carefully studied and analyzed so as to achieve a diversity of circular polarization (RHCP and LHCP) by switching the inputs.

The proposed antenna consisted of two symmetrical fractal structures as shown in **Figure 4**. The radiated plate is etched on a FR-4 substrate with $\varepsilon r = 4.3$, $\tan \delta = 0.027$, and compact size of $18 \times 18 \times 0.8$ mm^3, while the dimensions of the ground plate are 18 mm $\times$ 14.5 mm. **Figure 4** shows the initiator; it is a square patch with dimensions of ($L_o = W_o = 8.5$ mm). The square patch is modified by cutting equal slots ($0.1L_o = 0.1W_o$) in the middle of each arm and then modifying the dimensions of the arms at the corners by cutting increasable slips to change the dimensions of the arms at the corners in a way to configure asymmetrical corners, so the dimensions of each corner are not matching the others to generate different resonant frequencies that collected together to have wide impedance bandwidth. This modified square patch looks like the logo of Microsoft Office; it is the basic structure for the construction of the proposed antenna. All dimensions are illustrated in **Table 1**.

Figure 4 shows that antenna 0 (0th iteration) is made through the integration of two square patches. The first iteration consists of dual modified square patches, while the second iteration is made by adding a half size of the modified patch to the first iteration. The same procedure is to be applied to the third iteration, except for the fact that the additional modified patch has quarter size of the original one and antenna 4 (the fourth iteration) is configured by adding $\frac{1}{8}$ of the size of the original modified square patch to antenna 3. In this way, the modified patches are arranged in cascade arrangement so as to increase the total electrical length of the antenna of the same size. **Figure 4** represents the arrangement of the cascade modified patch (two symmetrical structures). According to Eq. (1), W_1 configures by adding a half size of the modified patch in **Figure 4** to both structures (left and right sides). The dotted yellow line in **Figure 5** is in the Y-axis and passes through all centers (black points) of the transformation structures (W_1, W_2 and W_3) which are calculated previously.

From Eq. (4), for the transformed function W_1, the scale factors are $r = s = 0.5$, the rotation angles $\theta = \varnothing = 0$, and the translation factors ($x_0 = 0, y_0 = 4.25$). Eq. (5) represents the values of scale factor ($r = s = 0.25$), the values of the rotation angle (0), and values of scale factor ($x_0 = 0, y_0 = 6.375$) for the transformed structure W_2. According to Eq. (6), the values of scale factor ($r = s = 0.125$), values of the rotation angle are (0), and values of scale factor are ($x_0 = 0, y_0 = 7.4375$) for the transformed structure W_3 .

So, antenna 1 (first iteration) is formed by the integration of two modified patch antennas—each is fed by symmetrical strip lines F_1 and F_2—the second iteration is made by adding half size of the modified patch to antenna 1 to configure antenna 2. The same procedure is applied in the third iteration except for the fact that the additional modified patch has quarter size of the original one (see **Figure 5**). Antenna 4 (fourth iteration) is configured by adding $\frac{1}{8}$ of the size of the original modified patch to antenna 3.

2.1 Resonant frequencies

The resonant frequency for the square patch antenna (the initiator) can be calculated by empirical Eq. (7), which is almost equal to ≈ 8.6 GHz. Since antenna 0 (0th iteration) is configured by the integration of twin square patches, there are two resonant frequencies, for one patch 8.5 GHz and for twin patch 4.3 GHz, which is matched with the notched resonant frequencies 4.5 GHz and 8.97 GHz observed in **Figure 6** (the dotted curve). By modifying the square patches in the first iteration

Para	mm	Para	mm	Para	mm	Para	mm	Para	mm	Para	mm
W	18	**Fw1**	1	**L1**	4.25	**W4**	2.9	**S6**	1.76	**D1**	0.7
L	18	**Fl2**	2	**W2**	3.9	**L4**	1.1	**S7**	1.5	**D2**	0.7
S1	2.9	**Fw2**	1	**L2**	3.5	**S3**	2.4	**S8**	1	**D3**	0.7
S2	2.9	**h**	0.8	**W3**	3.1	**S4**	1	**S9**	1.7	**D4**	0.7
Fl1	2	**W1**	4.3	**L3**	3.25	**S5**	2.1	**S10**	1.45		

Table 1.
Dimensions for the dual-input fractal antenna (mm).

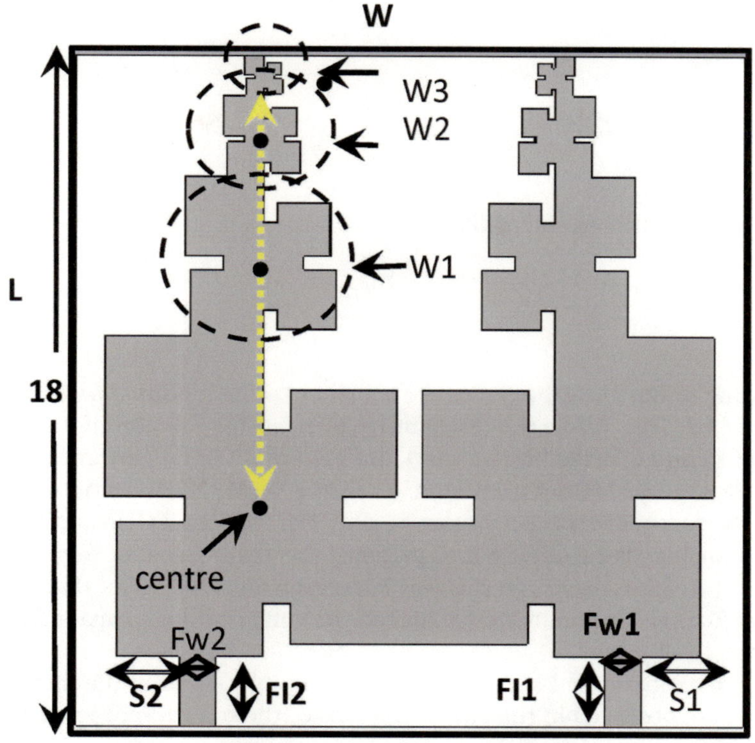

Figure 5.
The translation processes [17].

(antenna 1), the resonant frequency is shifted to 3.3 GHz, and the second band disappeared (the black curve):

$$f_r \approx \frac{c}{2L\sqrt{\varepsilon_r}} \tag{7}$$

The perimeter of the modified patch (**Figure 2c**) is 42 mm, while the perimeter of the square patch (**Figure 2a**) is 34 mm, and it is well known that different dimensions create different resonant frequencies. If the perimeter of the square patch is denoted by P0, the perimeter of the modified patch is P1, and the perimeter of the first iteration is P2, so:

$$P4 = \frac{1}{2}P3 = \frac{1}{4}P2 = \frac{1}{8}P1 \tag{8}$$

The perimeter of the cascade modified patches PC_i at each iteration can be calculated as:

$$PC_i = \frac{1}{2}PC_{i-1} + PC_{i-1}\, i = 1, 2, 3 \text{ and } 4 \tag{9}$$

The term (2L) in Eq. (7) represents the half perimeter of the square patch, so Eq. (7) can be rewritten as:

$$f_r \approx \frac{2C}{P\sqrt{\epsilon_{ef}}} \tag{10}$$

But, $\frac{\text{the perimeter of modified Patch}}{\text{the perimeter of square Patch}} = \frac{42}{34} = 1.23$
Eq. (10) will be:

$$f_i \approx \frac{2C}{1.23P_i\sqrt{\in_{ef}}} \tag{11}$$

Let $\frac{2C}{\sqrt{\in_{ef}}} = k$, and Eq. (11) will be:

$$f_i \approx \frac{k}{1.23PC_i} \tag{12}$$

According to Eq. (12), the resonant frequency of the modified patch (the first iteration) is 3.2 GHz. While the resonant frequencies for first iteration, which are presented in **Figure 6** (the black curve), are 3.3 and 7.1 GHz. However, these resonant frequencies are not useful for Wi-Fi and WiMAX applications. For second iteration, the resonant frequency calculated by Eq. (12) is 5.2 GHz and the lower resonant frequency is 2.6 GHz, which matched the lower resonant frequency shown in **Figure 6** (the blue curve). In this way, the resonant frequencies that are calculated using Eq. (12) approximately matched the values of the resonant frequencies, as presented in **Figure 6**.

Certain dual operating bands of 2.5–2.6 GHz and 5–6 GHz in the third iteration (antenna 3) are observed in the return loss curve (the red curve; **Figure 6**). The proposed antenna (fourth iteration) has a lower operating band of 2.4–2.6 GHz around resonant frequency 2.5 GHz and an upper operating band of 4.9–6 GHz around resonant frequency 5.1 GHz as shown in **Figure 6** (the dashed curve). These operating bands meet the specifications of Wi-Fi and WiMAX applications. So, the required operating bands are achieved by using novel fractal geometry at the fourth iteration in the same size of the 0th iteration.

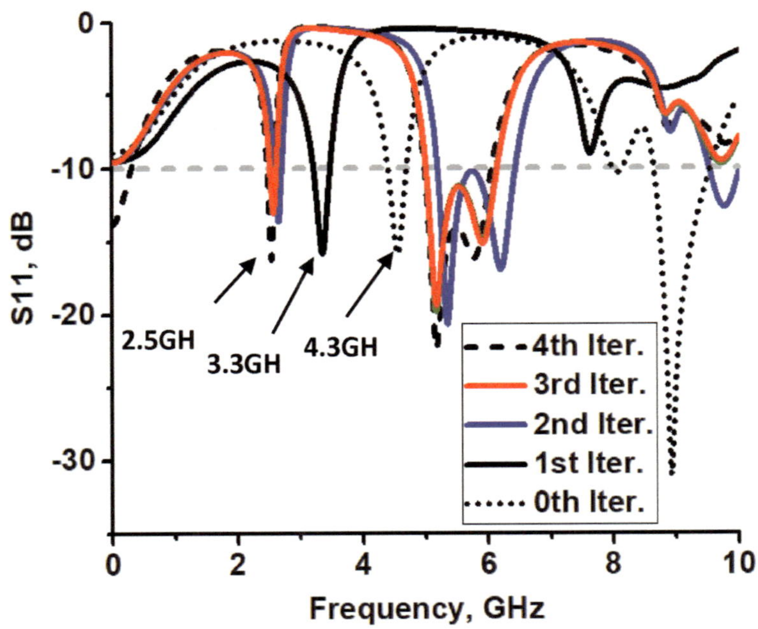

Figure 6.
Simulated return loss values for all antennas ($S_{11} = S_{22}$) [17].

2.2 Effect of the ground size

Figure 7 shows the effect of ground size on the impedance bandwidth of the proposed fractal antenna. Given that modifying the ground dimensions can change the antenna impedance and the matching factor between the antenna and the excitation port, the dimensions of the ground plate can be modified to select the operating band.

When the dimensions of the ground plate are set to 18 mm × 10 mm, the lower operating band shifts to 3 GHz and the upper operating band disappears as

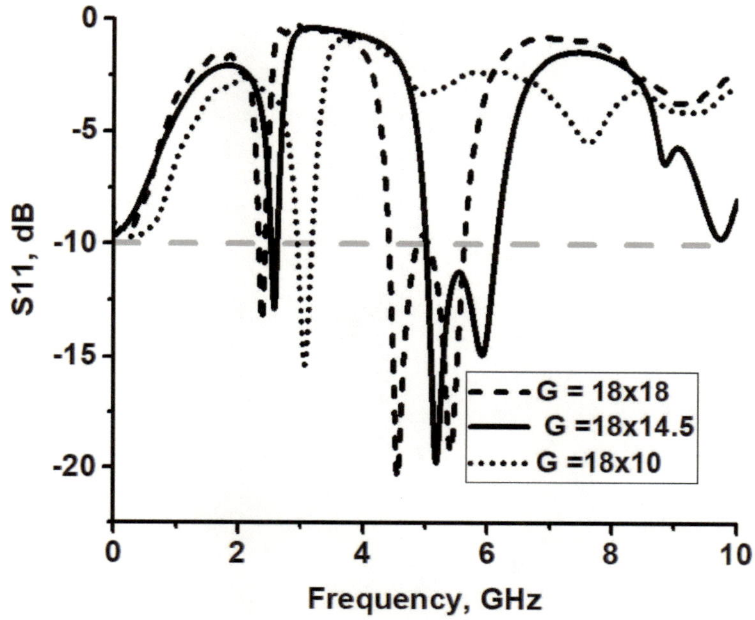

Figure 7.
The effect of ground plate size on reflection coefficient.

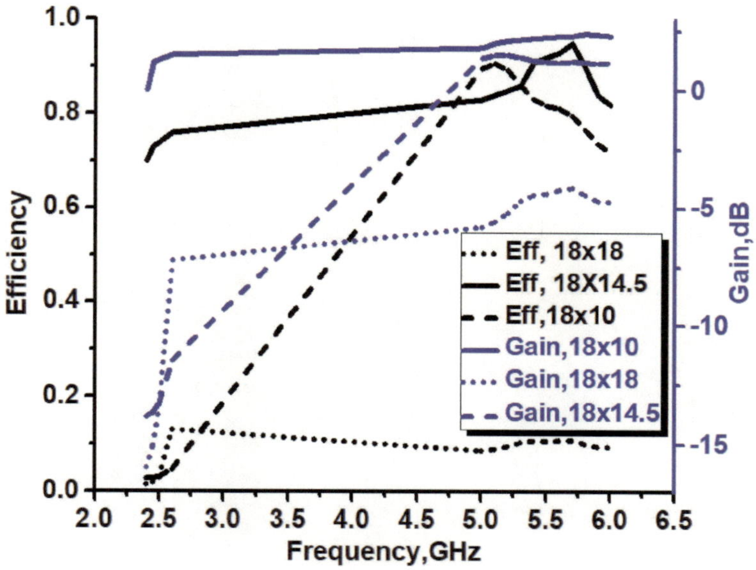

Figure 8.
The effect of ground plate size on the gain and the radiation efficiency [17].

Iteration	1st BW (2.4–2.6)	2nd BW (5–6)	Efficiency %	ARBW (GHz)	Gain (dB)	State
0th	(4–4.5)	(8.1–9.4)	40	—	−12 to 0	The resonant frequencies are out of the required, LP
1st	(3.1–3.6)	—	40–55	0.1	−8 to 0	2nd band disappeared, ARBW around 3.4 GHz
2nd	(2.6–2.7)	(5.4–6.4)	50–60	0.07	−2 to 1	Does not cover all required frequencies, ARBW around 5 GHz
3rd	(2.5–2.6)	(5–6)	55–68	0.2	−1.5 to 1.6	Does not cover all frequencies at the 1st band, ARBW around 3.5 GHz
4th	(2.4–2.6)	(4.9–6)	65–85	0.1, 0.3	0–2.4	Optimum

Table 2.
The radiation properties for all iterations of the dual-input fractal antenna.

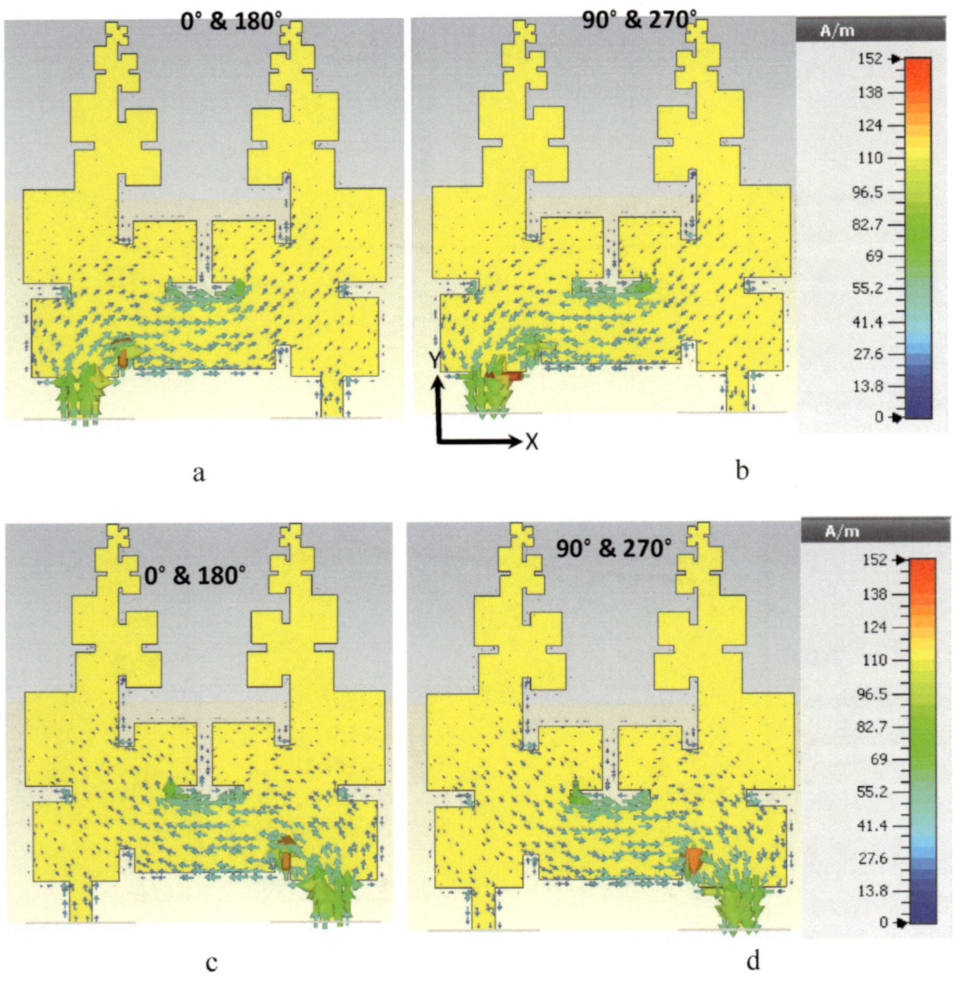

Figure 9.
Surface currents at 2.45 GHz for both inputs. (a) First input at 0° and 180° phase references. (b) First input at 90° and 270° phase references. (c) Second input at 0° and 180° phase references. (d) Second input at 90° and 270° phase references [17].

indicated by the dotted curve in **Figure 7**. The dashed curve in **Figure 7** represents the value of return losses when the dimensions of the ground plate are 18 mm × 18 mm, the lower band shifts to 2.3 GHz, while the upper band covers the 4.4–5.6 GHz frequency range. These bands are not suitable for Wi-Fi, WiMAX, and ISM applications. The solid curve in **Figure 7** indicates that the ground plate has optimum dimensions of 18 mm × 14.5 mm and that the dual notched bands meet the purpose of designing the fractal antenna.

Figure 8 presents the effect of ground size on the radiation efficiency (black curves) and gain (blue curves) of the proposed antenna. As shown in **Figure 8**, a fully grounded antenna (18 mm × 18 mm) has a low radiation efficiency (blue dotted curve) and gain (black dotted curve), especially at the lower operating band. The values of simulated gain and radiation efficiency are improved by reducing the size of the ground plate to 18 mm × 10 mm as indicated by the dashed curves in **Figure 8**. The gain increases to 0–1.5 dB at the lower operating band and to 1.5–2.5 dB at the upper operating band. Meanwhile, the radiation efficiency increases to 70–90% when the ground size is set to 18 mm × 14.5 mm as indicated by the black solid curve in **Figure 8**. Therefore, the optimum dimensions of the ground are 18 mm × 14.5 mm.

Table 2 illustrates some important radiation properties for each iteration, such as impedance bandwidth, efficiency, gain, and ARBW.

2.3 Surface current and circular polarization

Figure 9 shows the simulated surface current at the resonant frequency of 2.45 GHz for phase references of 0°, 90°, 180°, and 270°. When the first input for phase references at 0° and 180° is excited, most of the surface current direction on the feed strip line is along the +Y before circulating counter clockwise, as shown in **Figure 9a**.

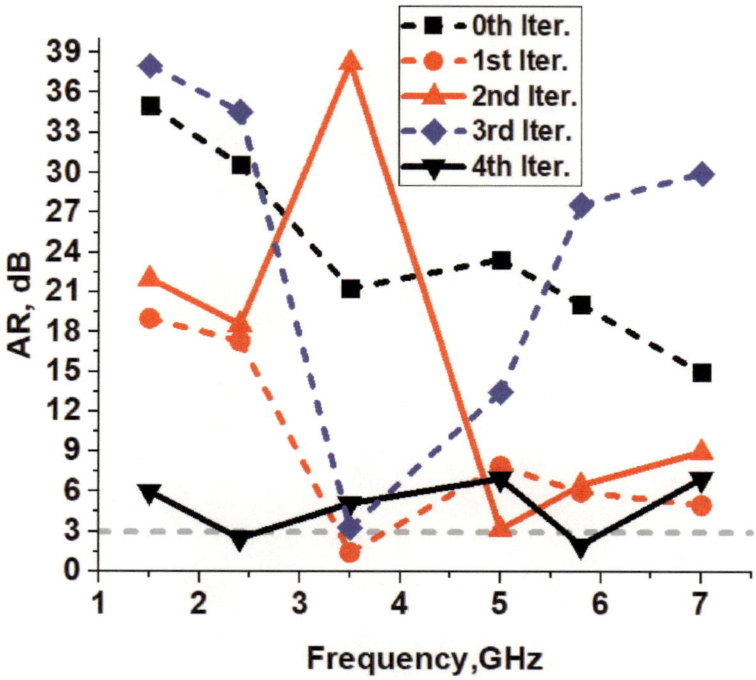

Figure 10.
Simulated axial ratio for all iterations [17].

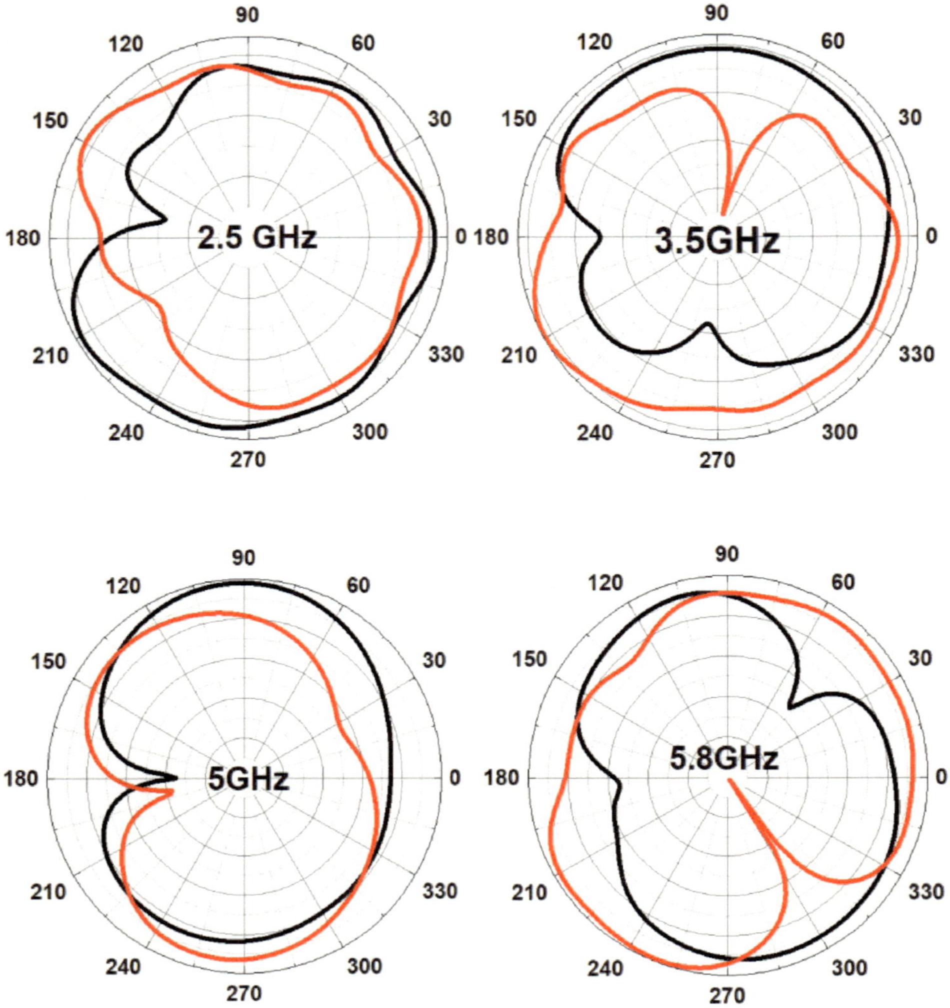

Figure 11.
Left (black curves) and right (red curves) polarization at 2.5, 3.5, 5, and 5.8 GHz [17].

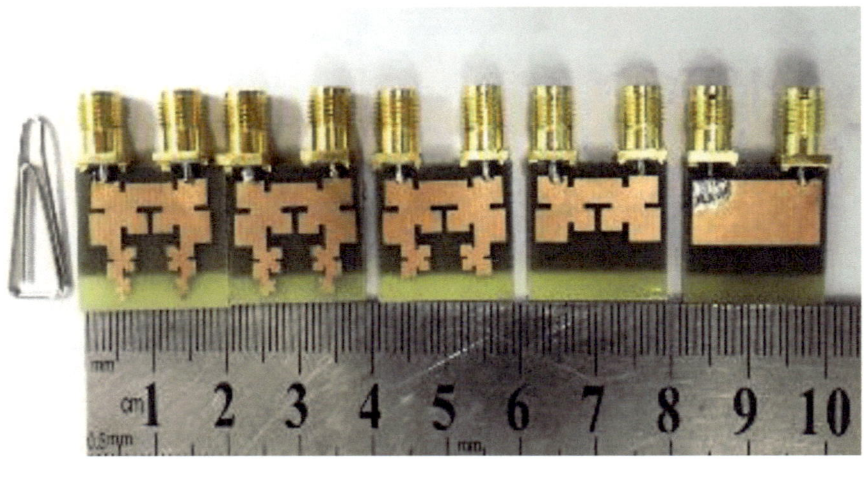

Figure 12.
The prototypes for all iterations [17].

For phase references at 90° and 270°, the surface current flows towards the −Y direction along the strip feeding line and then circulates clockwise, as shown in **Figure 9b**. Contrary to that, when the second input is excited, the surface current direction on the feed strip line and the circulating direction of the current changed, as shown in **Figure 9c** and **d**. RHCP and LHCP can be achieved by switching the two inputs. The signals of LHCP and RHCP can be received simultaneously. Thus, the proposed antenna has a dual circular polarization.

Figure 10 presents the AR values for all iterations. The AR values at the 0th iteration are too high, especially at the lower operating band as indicated by the dashed black curve. These values have changed during the configuration of the proposed antenna as shown in **Figure 10**. The ARBW of 0.09 GHz, which is approximately 66% of that of the lower operating band, is indicated by the solid black curve. Meanwhile, the ARBW of 0.35 GHz is approximately 30% of that of the upper operating band. The ARBW values in **Figure 10** match those that have been reported in previous circular polarization studies as shown in **Figure 10**.

Figure 11 shows the left and right polarizations at the frequencies of 2.5, 3.5, 5, and 5.8 GHz. The phase differences between the radiation patterns at 3.5 GHz and 5 GHz are 170° and 15°, respectively, whereas the LHCP and RHCP patterns at 2.5 GHz and 5.8 GHz shift by 88° (almost perpendicular to each other). Therefore, the antenna demonstrates circular polarization around the frequencies of 2.5 and 5.8 GHz. The antenna structure has zigzag edges that are configured by the arrangement of modified patches and creates lengthy paths for the surface current. At some frequencies, such as 2.5 and 5.8 GHz, the components of the surface currents are perpendicular to each other, thereby exciting orthogonal electric fields that, in turn, result in circular polarization [18]. The AR values in **Figure 10** and the current distribution in **Figure 9** match each other. The circular polarization is improved by the special design and arrangements of the modified structure during the antenna configuration, and the orthogonal components are generated at the resonant frequencies of 2.5 and 5.8 GHz.

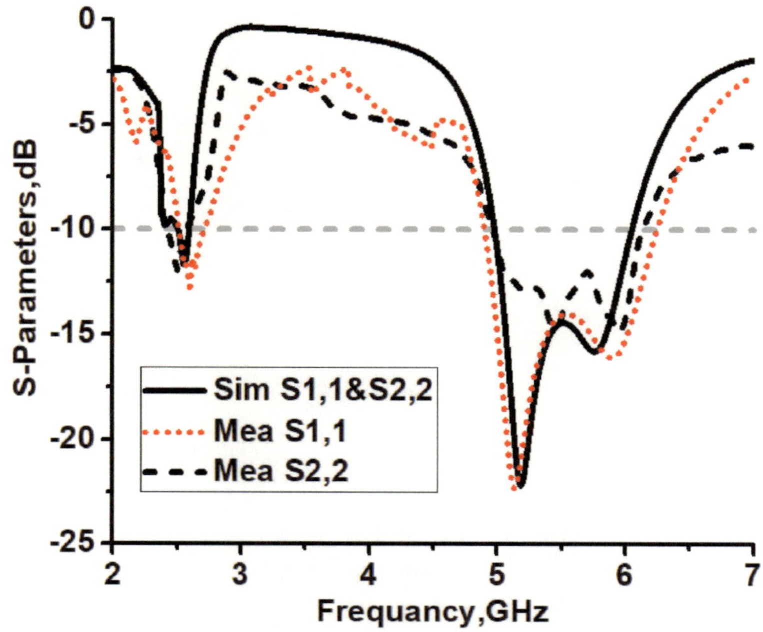

Figure 13.
The simulated and measured S-parameters for the proposed antenna [17].

2.4 Measurements and results

The five iterations are fabricated as shown in **Figure 12**. The compact size of the proposed fractal antenna is clear in this figure. The physical dimensions of the 0th iteration are the same of that for the fourth iteration, while the electrical length of the fourth iteration is much greater than the 0th iteration.

Figure 13 illustrates the simulated and measured reflection coefficients of the proposed fractal antenna. Generally, good matching is observed between simulated data (the solid curve) and the measured data (dashed and red dotted curves). However, several resonant frequencies in the experimental results are shifted unlike those in the simulated curve. Shifting occurs due to the impurity of materials used

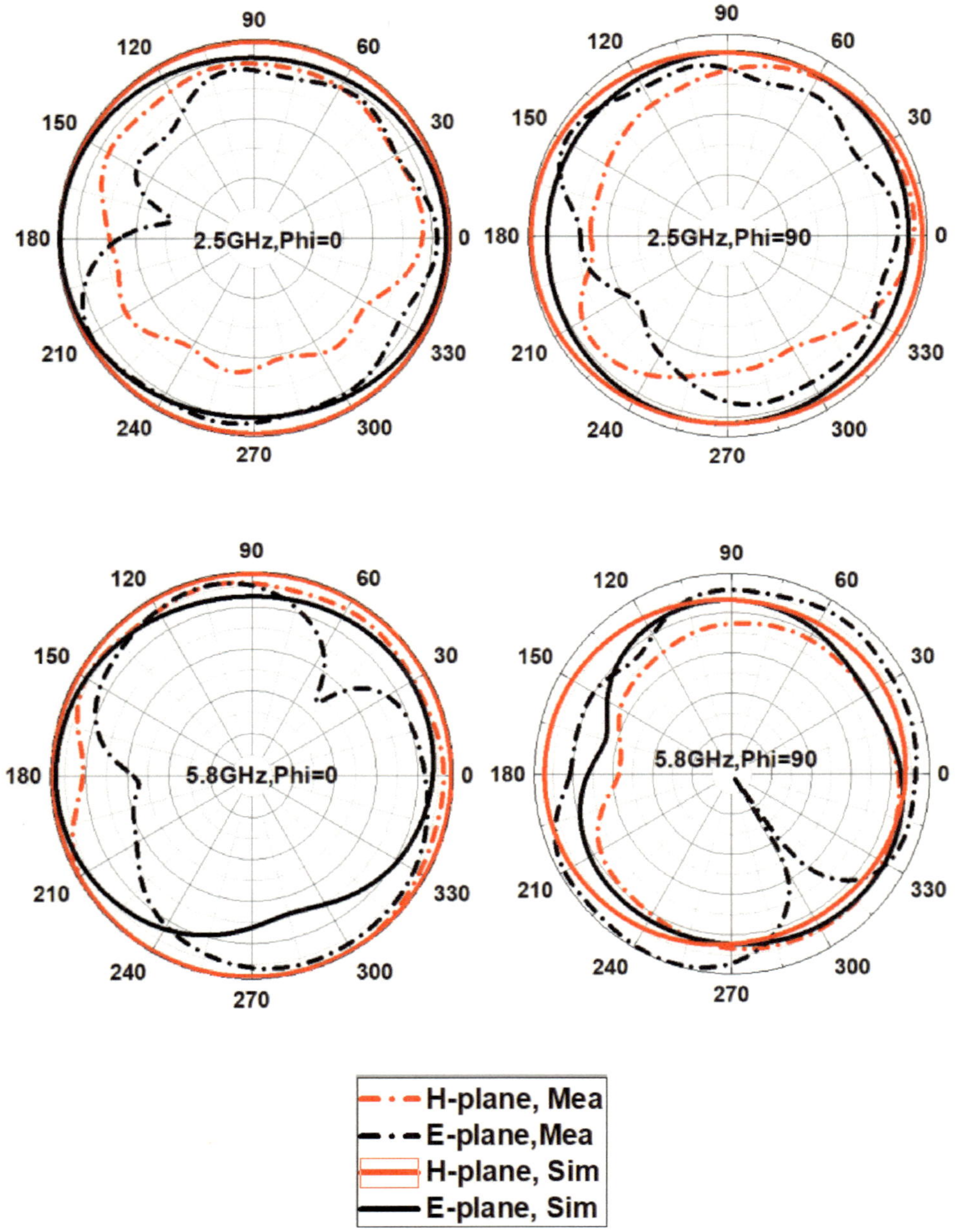

Figure 14.
Simulated and measured radiation patterns in E&H planes [17].

in the prototype. The impedance bandwidth of the first input (the red dotted curve) expanded to the range of 2.4–2.63 GHz and 4.8–6.4 GHz.

The measured reflection coefficient for the second input S_{22} is represented in **Figure 13** by the dashed curve. The measured impedance bandwidth expanded to the frequency band of 2.38–2.6 GHz and 4.9–6.1 GHz with low values of the reflection coefficient. Since the antenna is not MIMO antenna, the mutual coupling between the dual inputs S_{12} and S_{21} is not important in these measurements.

Figure 14 presents the simulated and measured radiation patterns on the E and H planes. The red or black solid curves represent the simulated data, while the red or black dashed curves represent the experimental results. The measured and simulated radiation curves in **Figure 14** show an acceptable agreement.

The measured radiation patterns on the H plane (the red dashed curve) at all resonant frequencies are almost omnidirectional. The E and H planes measured at 2.5 and 5.8 GHz when phi = 90 are perpendicular to each other, which matches the CP characteristics and the results of previous theoretical works on CP generation as shown in **Figures 11** and **12**. Meanwhile, the radiation pattern at the lower resonant frequency of 2.5 GHz is almost omnidirectional because the length of the surface current path is approximately half the wavelength of this frequency. Therefore, no side lobes are observed in its radiation pattern. As the frequency increases, the surface current path becomes greater than the wavelength of this frequency, thereby producing many side lobes at their radiation patterns. The radiation pattern is almost omnidirectional at resonant frequencies lower than 2.5 GHz because the length of the surface current path is approximately half the wavelength of this frequency. Therefore, side lobes are not observed in its radiation pattern. As the frequency increases, the surface current path becomes greater than the wavelength of this frequency, causing many side lobes at their radiation patterns.

The proposed fractal antenna displays an ARBW of 2.48–2.55 and 5.6–5.9 GHz, which is lesser than 3 dB as shown in **Figure 15** (the blue squared points) which is about 35% of the first operating band 2.4–2.6 GHz and about 30% of the second band 5–6 GHz. The values of the measured gain (the black circular points) vary

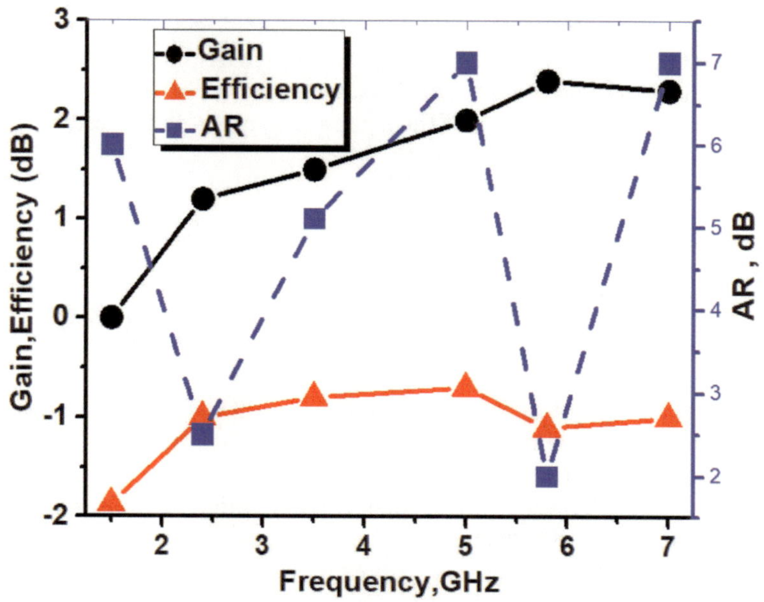

Figure 15.
The measured gain, efficiency, and AR [17].

between 0 dB at 2.5 GHz and 2.7 dB at 5.8 GHz. Meanwhile, the maximum efficiency (the red triangle points) is −0.7 dB (85%) at 5 GHz.

3. Meandered ring fractal antenna

The universal serial bus (USB) dongles are used in many portable communication devices such as laptops and pads in order to transmit and receive the data with a high bit rate. Since USB dongles are used with portable devices, the desired antenna must solve serious challenges such as the size, multiband, and the stability of radiation characteristics during the notched bands (gain and efficiency).

The aim of this study is to design a fractal ring antenna with a compact size and low profile, configured by three iterations, which covered the frequency range that meets the specification of the upper operating band for Wi-Fi and WIMAX applications and has high efficiency and stable radiation properties.

3.1 Antenna design

In this study, the proposed antenna with a compact size of $24 \times 9 \times 0.8\ mm^3$ was configured by a three-step process. The radiator of reference antenna (**Figure 16**) is a square patch with dimensions $x = y = 9$ mm and is separated from the ground by a gap of 0.3 mm. According to empirical Eq. (7), the dimension of the square patch will be equal to 36 mm for the resonant frequency (f_r = 5 GHz) and substrate FR-4 with $\varepsilon_r = 4.3$. To miniaturize the dimension of the antenna, let $X = Y = \frac{1}{4} \times 36\ \text{mm} = 9\ \text{mm}$. The ground plate features a rectangular shape with dimensions of $9.7 \times 9\ mm^2$. The radiator and ground are printed on the same side of commercial substrate (FR-4) with ε_r= 4.3, tan δ = 0.027, and thickness = 0.8 mm. **Figure 16** shows the first iteration in the design of antenna with square slot cuts in the radiator plate to configure the square ring with arm width equal to 0.1×. In the second iteration, the square ring is modified as a meandered ring to increase its electrical length to generate more resonant frequencies, which are collected to obtain a notched operating band of 4.4–6.7 GHz which is shown in **Figure 16** (the solid curve). All dimensions of the meandered ring are denoted as a function of x, which corresponds to the dimension of the square ring in the first iteration, as indicated in **Figure 16**. The length of the square ring in the first iteration equals 4x.

$$\text{Length of the upper arm} = 0.5X + 0.3X + 0.1X + 0.12X + 0.34X = 1.36X \quad (13)$$

$$\text{Length of left arm} = 0.34X + 0.17X + 0.1X + 0.22X + 0.38X = 1.21X \quad (14)$$

$$\text{Length of bottom arm} = 0.4X + 0.13X + 0.1X + 0.17X + 0.47X = 1.27X \quad (15)$$

$$\text{Length of right arm} = 0.44X + 0.1X + 0.1X + 0.13X + 0.52X = 1.29X \quad (16)$$

Thus, the total length of meandered ring equals 5.13X. The length of meandered ring increases by a factor of 1.28 compared with the length of the square ring of the same size in the first iteration. Of course, the increase in length of the four arms leads to increase in the length of the surface current paths and thus generates new resonant frequencies which collected together to give a wide impedance bandwidth, and this is clear in **Figure 17** (the solid curve) where the operating band increased when the ring becomes meandered at the second iteration.

Figure 17 depicts the real (black curves) and imaginary (blue curves) parts of the impedance values for the three antennas. The values of real-part impedance

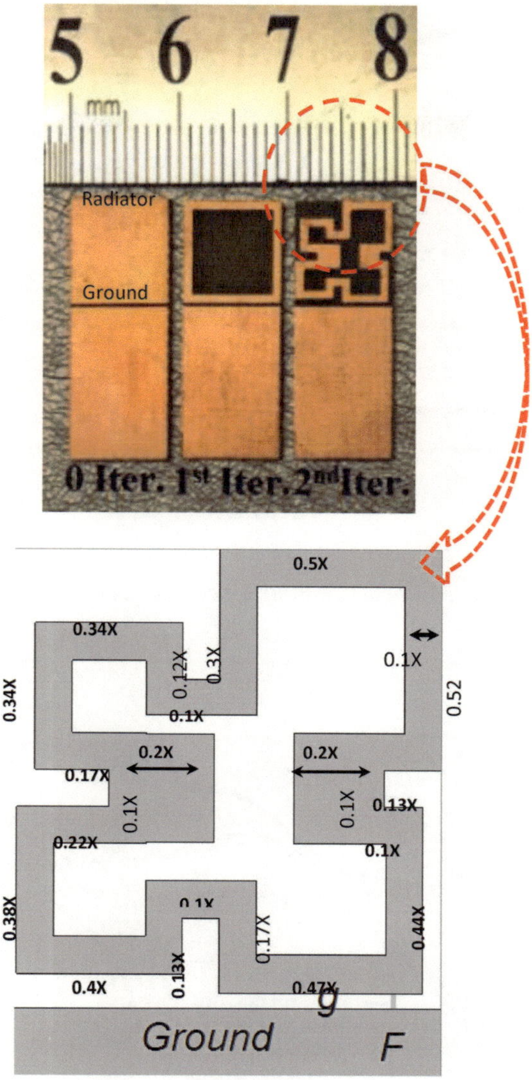

Figure 16.
The process of configuring the meandered ring antenna [19].

for all antennas are almost equal at frequencies < 4.7 GHz. Thereafter, the values of real-part impedance for the second iteration are closer to the input impedance of excitation port (50 ohms red line), especially in the frequency range of 5.1–6.4 GHz.

Figure 18 shows that negative imaginary values (capacitance) are observed in the curves of imaginary part impedance for all antennas at frequencies < 4.4 GHz. In the frequency band of 4.4–6 GHz, the values of the imaginary part are closer to zero (red line), that is, only real-part impedances for the three antennas are resistant. This property provides stable matching factor that leads to stable gain and efficiency in the operating band.

Table 3 illustrates some important radiation properties for each iteration, such as impedance bandwidth, efficiency, gain, and ARBW. It is clear that most of the required specifications that can be achieved at the second iteration are due to improving the values of radiation properties during the progress of antenna configuration, especially the impedance bandwidth, efficiency, gain, and an ARBW as shown in **Table 3**.

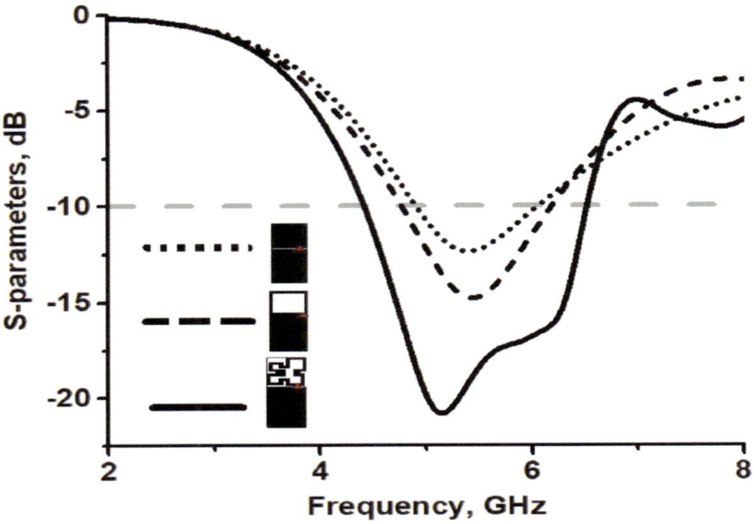

Figure 17.
The simulated reflection coefficient for the 0th iteration, first iteration, and the second iteration [19].

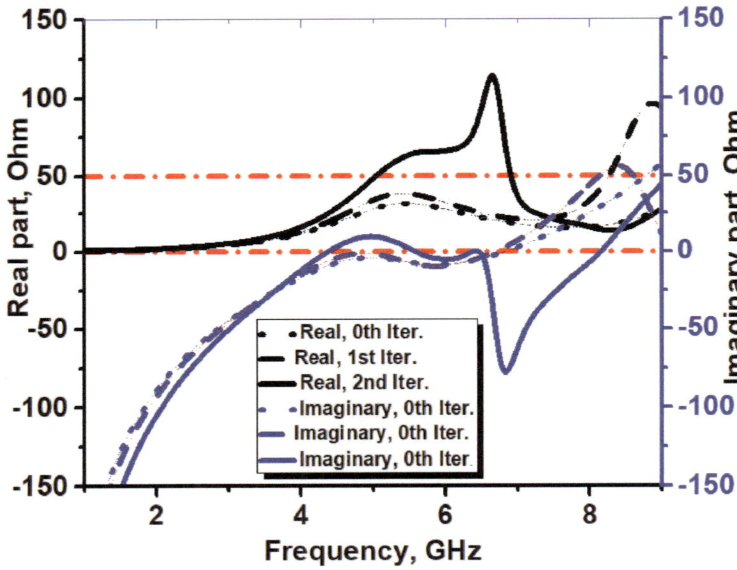

Figure 18.
The impedance values of the three iterations [19].

Iteration	BW (GHz)	Efficiency (%)	ARBW (GHz)	Gain (dB)	Lower S1, 1 (dB)	State
0th	(5.2–6)	50–65	—	−12 to 0	−10	Low efficiency, low gain, and LP
1st	(4.7–6)	60–75	—	1–3	−14	Does not have stable value of gain, LP
2nd	(4.4–6.7)	85–90	0.083	2.2–2.4	−20	Optimum

Table 3.
The radiation properties for all iterations of the meandered ring fractal antenna.

3.2 Current distribution

The maximum distribution currents at 5 GHz for the 0th iteration are mainly concentrated close to the feeding point on the patch and ground plates, as shown in **Figure 19**. In the first iteration, the currents are distributed in the additional area, especially on the square ring, which leads to the generation of new resonant frequency. The operating band is expanded compared with that at 0th iteration. The distribution of the surface current on ground plate is the same in all iterations because there is no change in surface current path at the ground plate during the progress of antenna configuration compared with the radiator plane.

Figure 19 (second iteration) shows that the surface current has two paths. The first path begins from the feeding point F and then passes through the right arm to point A, which has a total length equal to 1.79X = 16 mm, which is approximately one-fourth of the wavelength for the resonant frequency at 5 GHz.

The total length of the other path, which is almost perpendicular to the first and begins from feeding point F and then passes through the bottom arm to point B, is equal to 1.87X = 17 mm, which is approximately one-fourth of the wavelength for the resonant frequency at 5 GHz. Thus, two equal components of surface current normal to each other provide circular polarization radiation at 5 GHz.

3.3 Circular polarization

The length of the first surface current path from feeding point F to point A is 16 mm, which is almost equal to the length of the second surface current path from the feeding point to point B. These normal components are approximately ¼ of the wavelength for 5 GHz generated CP, as shown in **Figure 20**, which represents surface current distributions for phases at 0°, 90°, 180°, and 270° at 5 GHz resonant frequency. Surface current circulates clockwise along the upper and lower right quarters and counterclockwise along the lower left quarter for 0° and 180° phases. **Figure 20** shows the circulation direction of surface current when the phase becomes 90° and 270°.

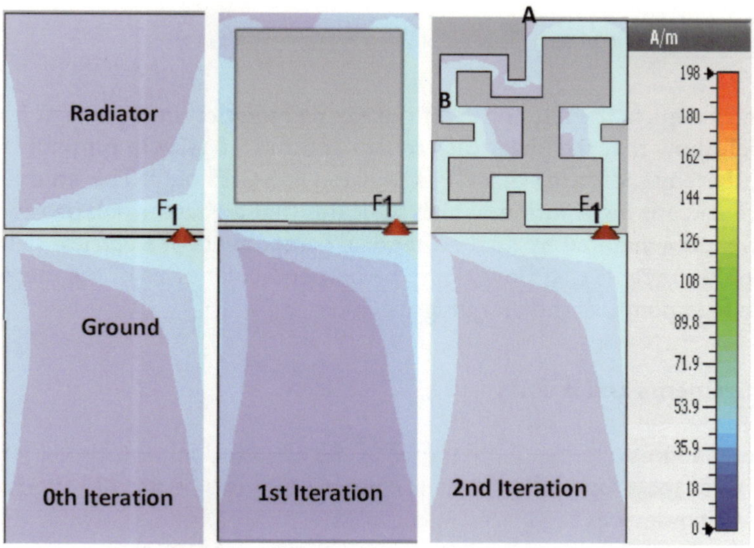

Figure 19.
The surface current at 5 GHz for all iterations [19].

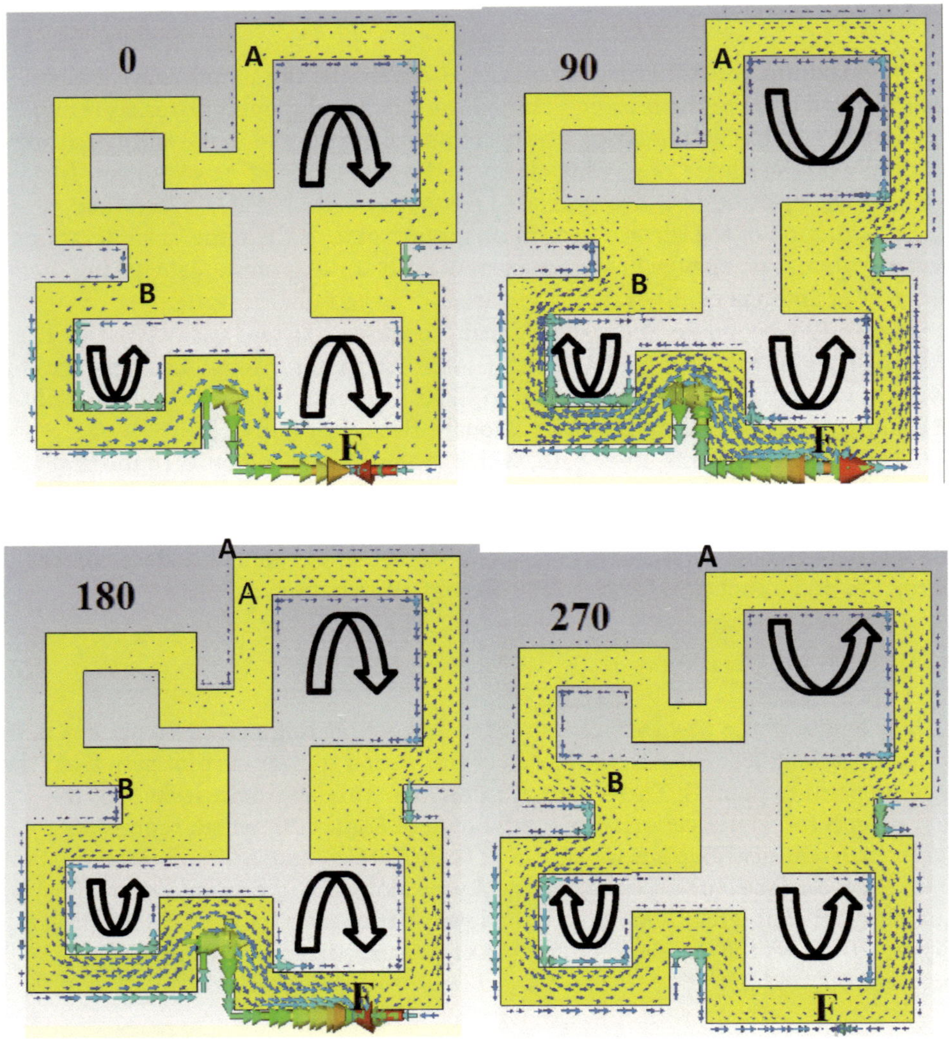

Figure 20.
Surface current for sequential phases (0°, 90°, 180°, and 270°) at 5 GHz [19].

Figure 21 depicts the left and right polarizing radiation in E-plane at 5 and 5.8 GHz, which show that the phase differences at 5 and 5.8 GHz is approximately 86° and 180°, thereby indicating that CP is radiated at 5 GHz only. This study matches the surface current distribution, which indicates that a circular polarization radiation at 5 GHz is generated by dual orthogonal components of electrical field created by the surface currents that flow along the perpendicular arms of the meandered ring at the feed point, as shown in **Figure 20**.

3.4 Measurements and results

Figure 22 represents the photographs of the all iteration prototypes; it is clear that the size of the proposed antenna is compact that can be used for portable communication devices.

The measured impedance bandwidth for the proposed antenna compresses to the frequency band of 4.8–6.7 GHz compared with the simulated impedance

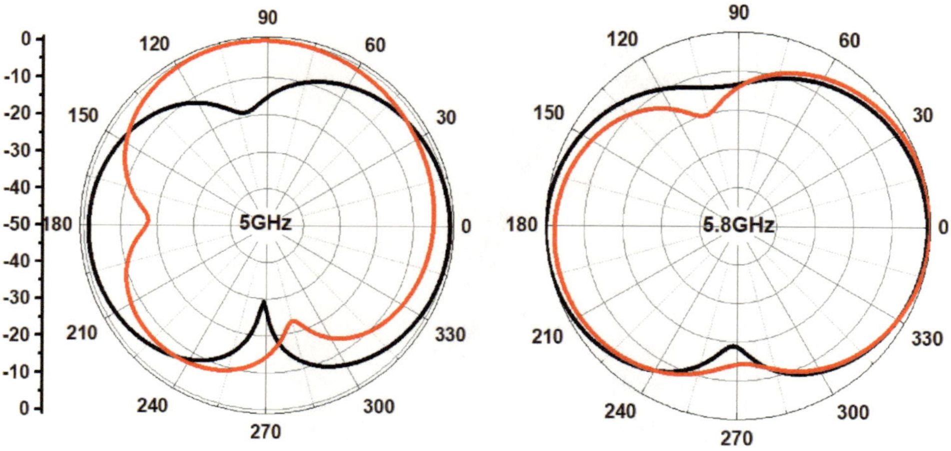

Figure 21.
Simulated left (black curves) and right (red curves) polarization in E-plane at 5 and 5.8 GHz [19].

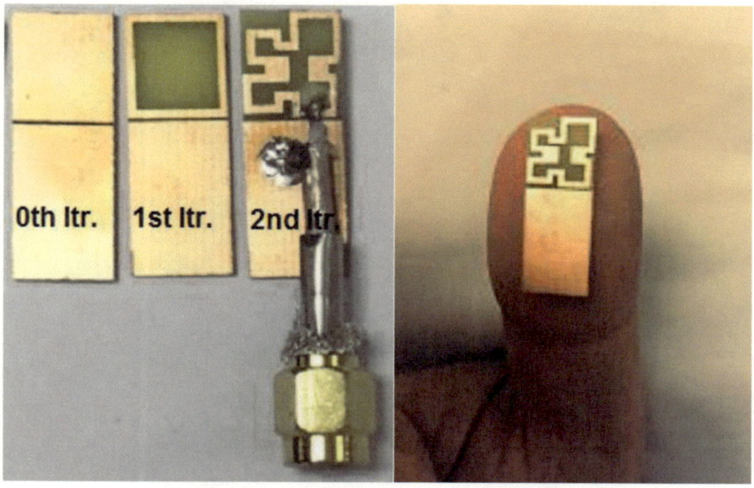

Figure 22.
The prototypes of all fractal antenna iterations [19].

bandwidth of 4.4–6.7 GHz, and the resonant frequency shifted to 5 GHz compared with 5.8 GHz for the simulated value as shown in **Figure 23**. That takes place due to impurities of some of the materials that are used in prototypes and due to the soldering.

Figure 24 shows a distinguishing agreement between simulated (solid) and measured (dashed) patterns. Radiation patterns in the H-plane (black curves) for the proposed antenna are almost omnidirectional at 5 and 5.8 GHz. **Figure 24b** shows that the radiation pattern at 5.8 GHz is similar to that at 5 GHz, but the former is more directional. At E-plane (when phi = 0), the radiation patterns looked like number 8 where two major lobes observed shifted by an angle of 180° as shown in **Figure 24a** (the red curves), while at 5.8 GHz, the radiation pattern in E-plane had dual asymmetrical major lobes.

Figure 25 depicts the measured values of the gain, efficiency, and an axial ratio. At frequency bands of 5–6 GHz, gain values are almost constant 2.3–2.4 dB, and efficiency reaches −0.45 dB (90%), whereas axial ratio bandwidth measures 83

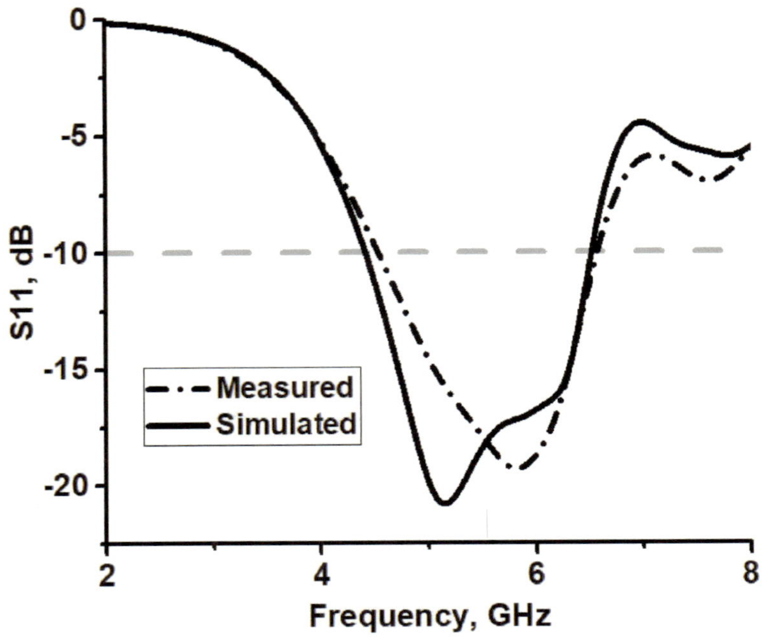

Figure 23.
Measured (dashed) and simulated (solid) $S_{1,1}$ for the proposed antenna [19].

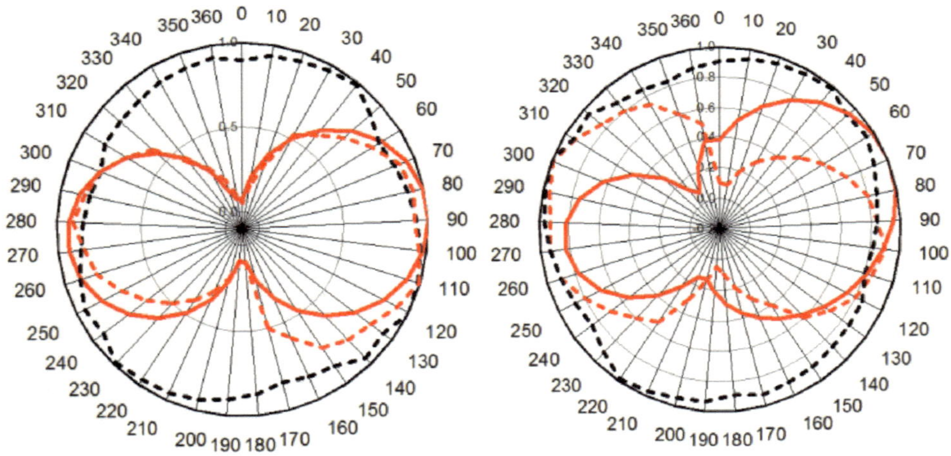

Figure 24.
Simulated (solid) and measured (dashed) radiation pattern. (a) At 5 GHz. (b) At 5.8 GHz [19].

MHz (approximately 8.3% of operating band) around a resonant frequency of 5 GHz.

Although the meandered ring fractal antenna has impedance bandwidth that covers only the upper band required for Wi-Fi, WiMAX, and ISM applications, it has stable radiation properties especially the gain and the efficiency. Furthermore, the meandered ring antenna has high efficiency (90%), highly compact size, and omnidirectional radiation pattern in H-plane that can be used for portable Wi-Fi and WiMAX devices.

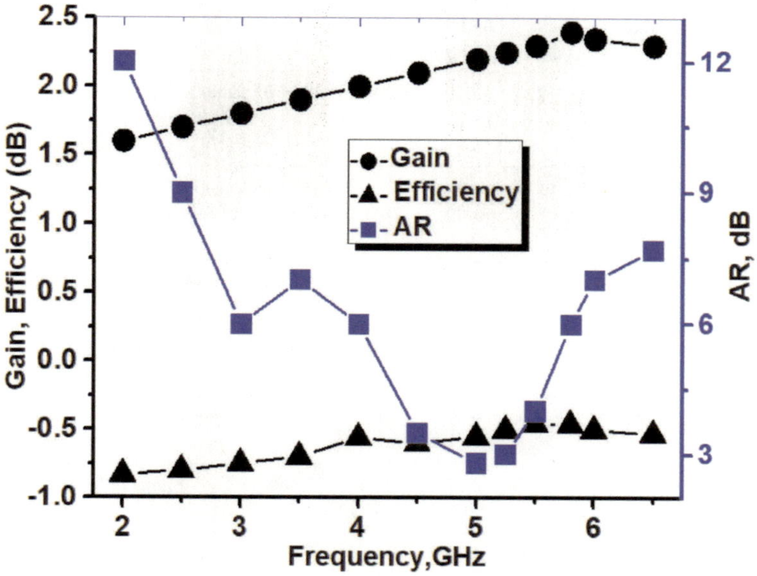

Figure 25.
Measured gain, efficiency, and AR for the antenna [19].

4. Conclusion

Fractal geometry is another type of micro strip antenna used in designing antennas. Two fractal antennas, namely, dual-input fractal antenna and meandered ring antenna, were investigated. All antennas are fabricated on commercial and cheap FR-4 substrate. The proposed fractal antennas are CP radiated by generating orthogonal components of electrical field.

The prototype of the dual-input fractal antenna has a compact size, which is approximately 9 and 16% of the Q-slot [20] and crescent slot [21] antenna size, respectively. The measured operating bands for the first input of the fractal antenna are 2.38–2.62 GHz and 5–6 GHz, whereas those for the second input of the fractal antenna are 2.4–2.65 and 4.8–6.2 GHz. The prototype of the dual-input fractal antenna displays an ARBW of 2.48–2.55 and 5.6–5.9 GHz, which is approximately 35% of the first operating band 2.4–2.6 GHz and approximately 30% of the second band 5–6 GHz. The values of the measured gain vary between 0 dBi at 2.5 GHz and 2.7 dBi at 5.8 GHz. The maximum efficiency is −0.7 dB (85%) at 5 GHz.

The fourth investigated antenna in this study is the meandered ring fractal antenna with compact size, which is approximately 9% of that of the Q-slot antenna, 10% of the crescent slot antenna size, and 66% of dual-input fractal antenna size. However, the meandered ring antenna only covers the upper bandwidth of 5–6 GHz used for Wi-Fi and WiMAX applications. The proposed antenna in this study has a small size, low profile, high measured efficiency (90%), stable gain (2.3–2.4) dBi, and CP radiation with an ARBW of 0.083 GHz, which is approximately 6% of the measured operating band at 4.8–6.2 GHz. Thus, the meandered ring monopole antenna is suitable for the requirements of portable communication devices.

Table 4 shows that the previous fractal antennas have circular polarization radiation, such as those reported in [3], which is the Giuseppe Peano fractal antenna that covered the frequency band of 1.5–4 GHz and ARBW of 0.2 GHz,

Ant.	BW (GHz)	Gain (dB)	ARBW (GHz)	Size (mm)	Effie. (%)	Weak points
[3]	(1.5–2.7)	3–4	0.2	60 × 60 × 10	40–80	Does not cover all required bands, low efficiency at lower frequency, used air gap
[6]	(2.32–2.52) (3.37–3.45) (5.6–5.9)	0.5 to –6	0.06, 0.09, 0.3	50 × 50 × 3.2	—	Used RT/duroid substrate, does not cover all frequencies in the upper band
[23]	(2.7–10.3)	—	1.1	54 × 36 × 1.5	85–95	No values of gain, missed 2.5 GHz band
[24]	(3.4–3.6)	6	0.1	40 × 40 × 1	91	Missed 2.4 GHz and 5 GHz bands
[22]	(2.2–2.6)	1.8t–4.6	0.6	100 × 100 × 13	—	Covered only 2.4 GHz band, large, no efficiency values
Fractal antenna	(2.4–2.6) (4.7–6.5)	0–2.4	0.07, 0.3	18 × 18 × 0.8	70–85	Missed 3.5 GHz band
Meander ring	(4.8–6.2)	2.3–2.4	0.083	24 × 9 × 0.8	85–90	Missed 2.4 GHz and 3.5 GHz bands

Table 4.
Comparisons between proposed fractal antennas and previous related works.

and the fractal antenna proposed in [6] whose size is approximately twice as that of the proposed fractal antenna is used in this study. The antenna has an ARBW of 0.06, 0.09, and 0.3 GHz at the three operating bands, but it does not cover all of the required frequencies, especially at the band of 5–6 GHz.

The meandered line fractal antenna is CP with an ARBW of 0.6 GHz reported in [22]. This meandered line antenna only covered the first band. The size of this antenna is very large and its efficiency has not been mentioned in the report. Modified Koch curve is used in designing the antenna based on Fibonacci sequence reported by [23]. The antenna is miniaturised to 54 × 36 × 1.5 mm and covered the frequency band (2.7–10.3) GHz with an ARBW of approximately 1.1 GHz around resonant frequency 6 GHz. The antenna does not cover the lower band (2.4–2.6) GHz, its CP radiation around 6 GHz, uses unbalanced feeding coaxial, and no gain values.

Therefore, the two proposed fractal antennas in the current study (dual fractal and meandered ring antennas) have specific characteristics better than the previous related antennas in **Table 4**. They are the best when possibly applied for Wi-Fi, WiMAX, and ISM communication purposes for the following reasons:

Firstly, the dual-input fractal antenna has dual operating bands for Wi-Fi (IEEE 802.11b,g,n) and WiMAX (IEEE 802.16e), Wi-Fi (IEEE802.11y), Wi-Fi (IEEE 802.11a,h,j,n), and WiMAX (IEEE 802.16d). The meandered ring fractal only covers the upper band (5–6) GHz.

Secondly, acceptable and greater values of gain for both antennas than the specified gain for fractal antennas in **Table 4** compared with their compact size.

Thirdly, the two types of fractal antennas are highly efficient (>80%), which matched the specific efficiency in **Table 4**. Despite the small size of the proposed antennas compared with many of the previous fractal antennas in **Table 4**, the manufactured antennas in the present study are characterized by superior and stable efficiency, thereby making them suitable for Wi-Fi and WiMAX applications.

Fourthly, the circular radiation properties at some operating bands for both proposed fractal antennas, that overcome the limits of generation CP with wide

ARBW in a compact size, especially the dual fractal antenna, which has CP radiation during the first and third bands, can avoid many serious problems, such as mismatched polarization and multipath interferences.

Fifthly, the highly compact size and omnidirectional radiation patterns in H-plane, especially the meandered ring antenna, facilitated the possibility of using the proposed antennas for mobile Wi-Fi and WiMAX communication devices.

Sixthly, aside from being low profile and easy to fabricate and the use of commercial FR-4 substrate in addition to existing features, these antennas can be adopted for the purpose for which they are designed.

Author details

Amer T. Abed[1]*, Mahmood J. Abu-AlShaer[2] and Aqeel M. Jawad[2]

1 Communication Engineering Department, Al-Mamun University College, Baghdad, Iraq

2 Al-Rafidain University College, Filastin, Baghdad, Iraq

*Address all correspondence to: amer.t.abed@ieee.org

References

[1] Werner DH, Gangul S. An overview of fractal antenna engineering research. IEEE Anlennas and Propagation Magazine. 2003;**45**:38-57

[2] Chen W-L, Wang G-M, Zhang C-X. Bandwidth enhancement of a microstrip-line-fed printed wide-slot antenna with a fractal-shaped slot. IEEE Transactions on Antennas and Propagation. 2009;**57**:2176-2179

[3] Oraizi H, Hedayati S. Miniaturization of microstrip antennas by the novel application of the Giuseppe Peano fractal geometries. IEEE Transactions on Antennas and Propagation. 2012;**60**: 3559-3567

[4] Chaimool S, Chokchai C, Akkaraekthalin P. Multiband loaded fractal loop monopole antenna for USB dongle applications. Electronics Letters. 2012;**48**:1446-1447

[5] Srivatsun G, Subha Rani S. Compact multiband planar fractal cantor antenna for wireless applications: An approach. International Journal of Antennas and Propagation. 2012;**2012**

[6] Reddy V, Sarma N. Triband circularly polarized Koch fractal boundary microstrip antenna. IEEE Antennas and Wireless Propagation Letters. 2014;**13**:1057-1060

[7] Tripathi S, Mohan A, Yadav S. Hexagonal fractal ultra-wideband antenna using Koch geometry with bandwidth enhancement. IET Microwaves, Antennas & Propagation. 2014;**8**:1445-1450

[8] Weng W-C, Hung C-L. An H-fractal antenna for multiband applications. IEEE Antennas and Wireless Propagation Letters. 2014;**13**:1705-1708

[9] Dhar S, Patra K, Ghatak R, Gupta B, Poddar DR. A dielectric resonator-loaded minkowski fractal-shaped slot loop heptaband antenna. IEEE Transactions on Antennas and Propagation. 2015;**63**:1521-1529

[10] Li D, Mao J-F. Coplanar waveguide-fed Koch-like sided Sierpinski hexagonal carpet multifractal monopole antenna. IET Microwaves, Antennas & Propagation. 2014;**8**:358-366

[11] Yogesh Kumar Choukiker SKB. Wideband frequency reconfigurable Koch snowflake fractal antenna. IET Microwaves, Antennas & Propagation. 2017;**11**:203-208

[12] Abed AT, Singh MS, Islam MT. Amer fractal slot antenna with quad operating bands high efficiency for wireless communications. In: 2016 IEEE 3rd International Symposium on Telecommunication Technologies (ISTT). 2016. pp. 6-8

[13] Abed AT. Novel sunflower MIMO fractal antenna with low mutual coupling and dual wide operating bands. International Journal of Microwave and Wireless Technologies. 2020

[14] Abed AT. A new fractal antenna flame structure with impedance ration 10.6: 1 high efficiency for UWB applications. Al-Ma'mon College Journal. 2017:215-231

[15] Abed AT. A novel coplanar antenna butterfly structure for portable communication devices, a compact antenna with multioperating bands. IEEE Antenna and Propagation Magazine. 2020

[16] Werner DH, Haupt RL, Werner PL. Fractal antenna engineering: The theory and design of fractal antenna arrays. IEEE Antennas and Propagation Magazine. 1999;**41**:37-58

[17] Amer MSJS, Abed T, Islam MT. Compact fractal antenna circularly

polarised radiation for Wi-Fi and WiMAX communications. IET Microwaves, Antennas & Propagation. 2018;**12**:2218-2224

[18] Abed AT, Singh MSJ, Jawad AM. Investigation of circular polarization technique in Q-slot antenna. International Journal of Microwave and Wireless Technologies. 2020;**12**:176-182

[19] Abed AT, Singh MS, Islam MT. Compact-size fractal antenna with stable radiation properties for Wi-Fi and WiMAX communications. KSII Transactions on Internet & Information Systems. 2018;**12**

[20] Abed A, Singh M. Slot antenna single layer fed by step impedance strip line for Wi-Fi and Wi-Max applications. Electronics Letters. 2016;**52**:1196-1198

[21] Abed AT, Singh MSJ, Islam MT, Khaleel AD. Dual crescent-shaped slot antenna fed by circular polarisation into dual orthogonal strip lines. IET Microwaves, Antennas & Propagation. 2017;**11**:2129-2133

[22] Luo K, Chen B, Ding W-P. Meander line coupled cavity-backed slot antenna for broadband circular polarization. IEEE Antennas and Wireless Propagation Letters. 2015;**14**:1215-1218

[23] Chetna Sharma SM, Vishwakarma DK. Miniaturization of spiral antenna based on fibonacci sequence using modified Koch curve. IEEE Antennas and Wireless Propagation Letters. 2017; **16**:932-935

[24] Cai T, Wang G-M, Zhang X-F, Shi J-P. Low-profile compact circularly-polarized antenna based on fractal metasurface and fractal resonator. IEEE Antennas and Wireless Propagation Letters. 2015;**14**:1072-1076

Chapter 5

Radiation Pattern Synthesis of Planar Arrays Using Parasitic Patches Fed by a Small Number of Active Elements

Jafar Ramadhan Mohammed and Karam Mudhafar Younus

Abstract

In this chapter, several planar array designs based on the use of a small number of the active elements located at the center of the planar array surrounded by another number of the uniformly distributed parasitic elements are investigated. The parasitic elements are used to modify the radiation pattern of the central active elements. The overall radiation pattern of the resulting planar array with a small number of active elements is found to be comparable to that of the fully active array elements with a smaller sidelobe level (SLL) at the cost of a relatively wider beamwidth and lower directivity. Nevertheless, the uses of only a small number of the active elements provide a very simple feeding network that reduces the cost and the complexity of the array. Simulation results which have been computed using computer simulation technology-microwave studio (CST-MWS) show that the sidelobe level of the designed array pattern with parasitic elements is considerably better than that of the similar fully active array elements. The proposed array can be effectively and efficiently used in the applications that require wider antenna beams.

Keywords: planar arrays, parasitic elements, driven elements, mutual coupling, feeding network, array pattern synthesis

1. Introduction

By properly controlling the design parameters of the antenna arrays, it is possible to generate the desired radiation characteristics such as higher directivity and lower sidelobe level. The most design parameters that have been given an increased attention for many practical applications are the amplitude and the phase excitations of the individual elements [1–3] in addition to the element's spacing [4, 5]. However, the practical implementation of the feeding networks in such arrays is known to be very complex and expensive.

Recently, the feeding networks were made very easy and practically efficient by controlling only a certain number of the active elements rather than all of the array elements [6, 7]. Other methods used a number of parasitic elements illuminated by a single active element to simplify the complexity issue of the feeding network.

These methods are of special interest because they are capable of introducing extra degrees of freedom that could effectively contribute to synthesize the desired array patterns without making any modification in the feeding networks of the active elements [8, 9]. Particularly, in [8], a planar array of complete parasitic dipoles which was placed under a single active λ/2-dipole element to modify its radiation pattern was studied, while in [9], the authors considered a circular array of parasitic dipoles centered by one active dipole. The desired pattern was obtained by properly changing the spacing between the parasitic elements and the length of the parasitic elements, while the position and length of the driven element were kept fixed.

This chapter presents a new method for array pattern configuration with a low sidelobe level. The element type of the considered planar array is designed to be in the form of a simple rectangular patch. The feeding points of the active patches are optimally chosen such that the value of the VSWR becomes minimum. Moreover, the location of the feeding point has been obtained by the use of an optimization algorithm supported by the CST-MWS. In the proposed array, the elements are divided into two sub-arrays. The elements of the first sub-array which are located on the center of the planar array is referred as an active array, which contains only a small number of the array elements (usually one active element for an array of 3 × 3 elements and four active elements for other array sizes), whereas the second array subset contains the majority of the array elements which are parasitic, and their distances from the active ones need to be carefully chosen. Moreover, the second array subset may be selected either as a single or multiple rectangular rings according to the design requirements. The CST software is used to choose the most effective distances between the active and the parasitic elements (i.e., the most appropriate perimeter of the outer rectangular ring) such that the greatest reduction in the sidelobe level could be obtained with a smaller scarifies in the directivity.

2. Problem formulation

In order to highlight the major contribution of the proposed work, first we plot the radiation pattern of the fully active planar array elements that are designed with nonuniform amplitude excitation according to the Dolph polynomial to get the desired sidelobe level. For comparison, the uniformly excited planar array with 6 × 6 active elements is also included. **Figure 1** shows the radiation patterns of the Dolph excited planar array with SLL = −40 dBi and the uniformly excited planar

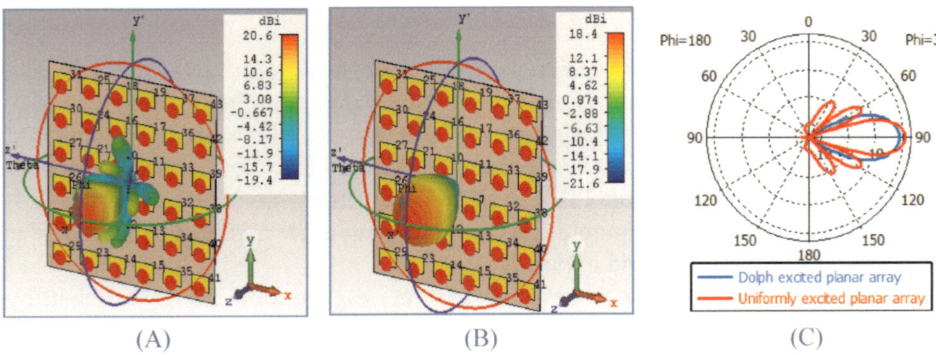

Figure 1.
Results for 6 × 6 fully active array elements: (A) uniformly excited planar array; (B) Dolph excited planar array with SLL = −40 dBi; (C) radiation patterns in polar plots.

	Fully active elements with Dolph excitations					Proposed array with parasitic elements				
Element	Amplitude	Phase	SLL (dB)	HPBW	Directivity (dB)	Amplitude	Phase	SLL (dB)	HPBW	Directivity (dB)
±(1.1)	0.0402	0	−21.4	23.1	18.4	0	0	−18.7	33.6	14.8
±(1.2)	0.1240	0				0	0			
±(1.3)	0.2004	0				0	0			
±(2.1)	0.1240	0				0	0			
±(2.2)	0.4675	0				0	0			
±(2.3)	0.6184	0				0	0			
±(3.1)	0.2004	0				0	0			
±(3.2)	0. 6184	0				0	0			
±(3.3)	1.0000	0				1	0			
±(4.1)	0.2004	0				0	0			
±(4.2)	0.6184	0				0	0			
±(4.3)	1.0000	0				1	0			
±(5.1)	0.1240	0				0	0			
±(5, 2)	0.4675	0				0	0			
±(5.3)	0.6837	0				0	0			
±(6.1)	0.0874	0				0	0			
±(6.2)	0.2021	0				0	0			
±(6.3)	0.2956	0				0	0			

Table 1.
The amplitude excitations of the Dolph and the proposed arrays.

array. In this figure, both array sizes are chosen to be 6 × 6 active elements, and the separation distance between any two successive elements in X- or Y-axis is chosen to be λ/2. **Table 1** shows the corresponding element excitations for the nonuniformly Dolph-distributed array.

From **Figure 1**, it can be seen that the Dolph excited array has a pattern with wider beamwidth and lower directivity than that of the uniformly excited planar array. More important, it can be seen that the Dolph excited array needs to be designed with a very precise values of the amplitude excitations to get the desired sidelobe pattern. Any simple variations from those computed values will lead to significant changes in the corresponding radiation pattern. In practice, the main reason for such variations in the amplitude excitations is due to the use of digital attenuators which are associated with the quantization errors due to the limited number of the quantized levels. Thus, the actual patterns are usually different from the desired ones.

In order to solve such an important problem in practice, the authors suggest an alternative method for obtaining the desired radiation pattern that does not need any digital RF components for amplitude scaling. In the proposed array, the magnitude excitations of the central active elements are assigned to the values of 1, while the parasitic elements are assigned to the values of 0. As the number of the active elements in the proposed array is chosen to be small, the value of the obtained gain is expected to be lower than that of the fully active array elements.

Moreover, the effect of the scattering, the mutual coupling, and the element type is all considered with the proposed array. Thus, the desired and the actual patterns are both expected to be in a good agreement.

Figure 2 shows the results of applying the proposed method under the use of only 4 active elements out of 36 elements (i.e., the number of the parasitic elements is 32). As can be seen, the proposed method with only four active elements is still capable to provide an acceptable radiation pattern as compared to that array of fully active elements.

As mentioned, the corresponding amplitude excitations of all of the abovementioned arrays are shown in **Table 1**.

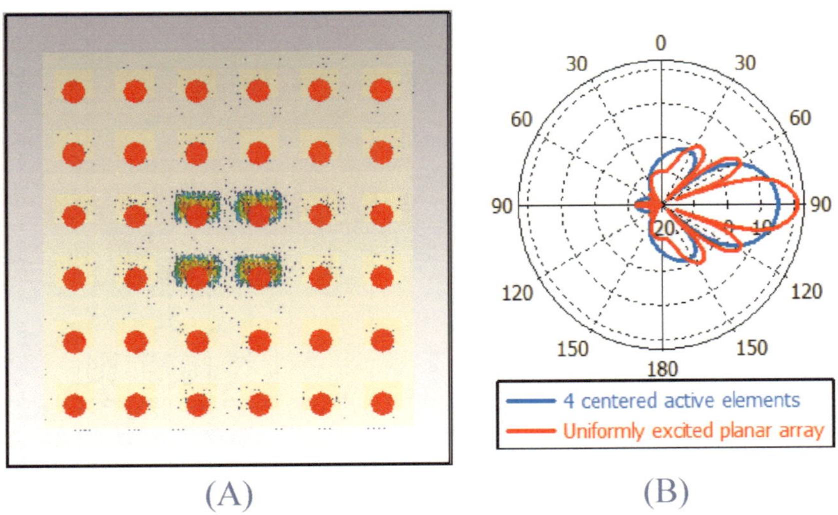

Figure 2.
Results for the proposed planar array with 4 centered active elements and 32 parasitic elements; (A) current distribution and (B) radiation patterns in polar plots.

3. Description of the designed array

In this section, the procedures of designing a single patch element are presented; then it is straight forwarded to construct a required two-dimensional planar array. Generally, there are three methods in the design of a single microstrip patch: the transmission line model, cavity model, and full-wave model [1]. Among them, the transmission line model is considered in this work. This is mainly due to its simplicity and popularity. First, the patch dimensions (width and length) are designed according to the following:

$$\text{Width} = \frac{1}{2\,f_r\sqrt{\mu_0\epsilon_0}}\sqrt{\frac{2}{\epsilon_r+1}} = \frac{v_0}{2\,f_r}\sqrt{\frac{2}{\epsilon_r+1}} \tag{1}$$

where v_0 is the velocity of the electromagnetic wave in the dielectric and f_r is the resonant frequency. On the other hand, the patch length usually seems to be electrically greater than its physical dimension due to the fringing effects. Consequently, the effective length (L_{eff}), effective dielectric constant (ϵ_{eff}), and the length extension (ΔL) have been computed as given by

$$\epsilon_{eff} = \frac{\epsilon_r+1}{2} + \frac{\epsilon_r-1}{2}\left[1+\frac{12h}{WP}\right]^{-1/2} \tag{2}$$

$$L_{eff} = \frac{c}{2\,f_r\sqrt{\epsilon_{eff}}} \tag{3}$$

$$\Delta L = 0.412h\,\frac{(\epsilon_{eff}+0.3)\left[\frac{WP}{h}+0.264\right]}{(\epsilon_{eff}-0.258)\left[\frac{WP}{h}+0.813\right]} \tag{4}$$

Then, the accurate length value of the patch is computed from

$$\text{Length} = L_{eff} - 2\Delta L = \frac{c}{2f_r\sqrt{\epsilon_{eff}}} - 2\Delta L \tag{5}$$

Figure 3 shows a single patch element.

Note that, in this work, the coaxial probe is used to feed the single patch element. It is designed such that the value of reflection coefficient (S11) is minimized at selected frequency value 2.4 GHz as shown in **Figure 4**.

Finally, to get the required planar array, a number of the designed patches are uniformly distributed along the X- and Y-axis. The radiation pattern of such array is then obtained by the vector addition of the radiation pattern of each individual

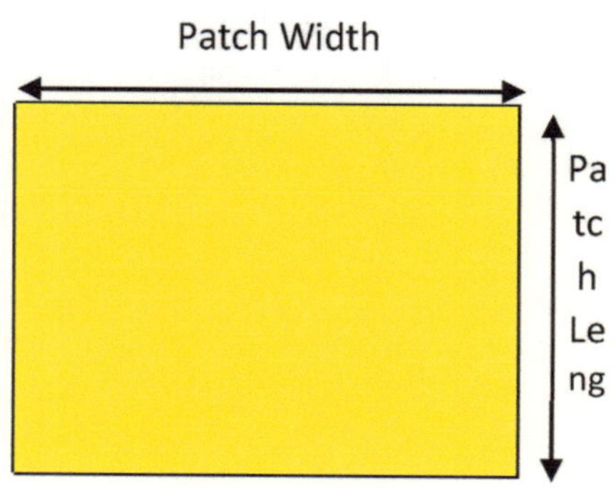

Figure 3.
Single patch element.

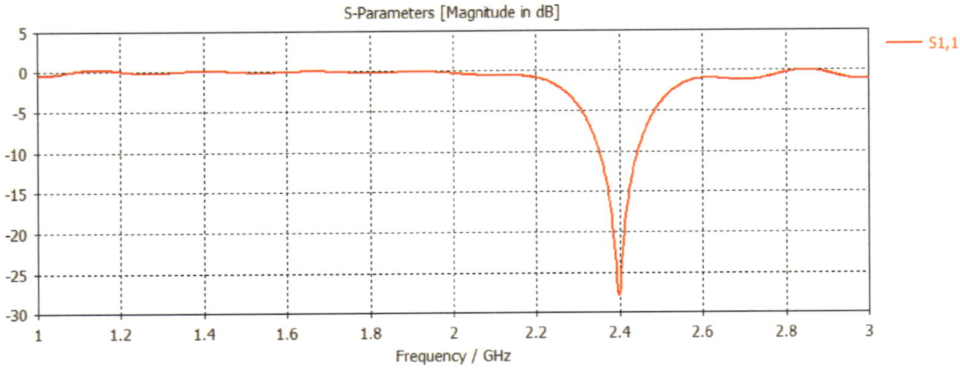

Figure 4.
Reflection coefficient (S11) of the proposed patch antenna.

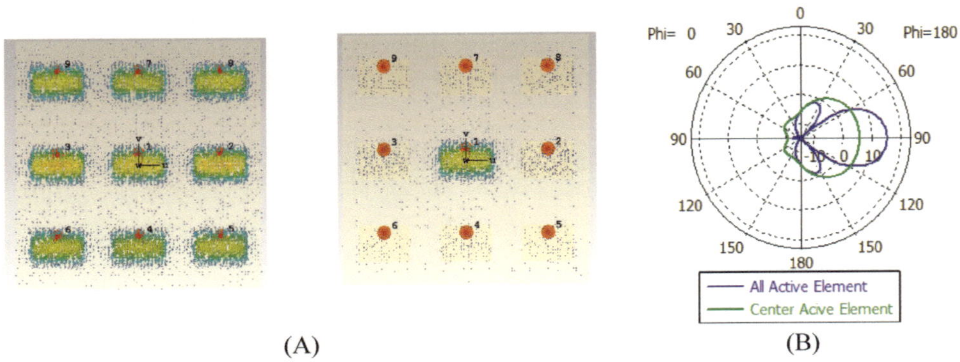

Figure 5.
Results for 3 × 3 planar arrays: (A) current distributions on both active and parasitic elements and (B) radiation patterns in the polar plot.

patch, taking into account the effect of mutual coupling between the array elements and the scattering effect.

4. Simulation results

To illustrate the effectiveness of the proposed method, different array configurations have been considered. In all examples, a rectangular planar array has been considered with $(N \times M)$ elements distributed on the XY-plane. The operating wavelength is chosen to be λ = 125 mm, and the separation distance between any two adjacent patches in both X- and Y-axis is d = $\lambda/2$ = 62.5 mm. The target angle is assumed to be at $(\theta, \varphi) = 0^o$.

In the first example, an array with 3 × 3 elements has been considered, and only the center element is chosen to be active, while the others are set to be parasitic. **Figure 5** shows the current distributions on the active and parasitic elements for both the fully active elements and the proposed array with only a single active element at the center of the array. For the fully active elements, the directivity (D) was 14.85 dBi, the sidelobe level was −16.69 dBi, and the half-power beamwidth (HPBW) was 33.6°, whereas for the proposed array these values were found to be D = 5.41 dBi, SLL = −13.22 dBi, and HPBW = 97.30°.

In the second example, a planar array with 4 × 4 elements has been considered, and the separation distances between any two elements on the X- and Y-axis was set to λ/2. Here, for the proposed array, only four centered elements are chosen to be active, while the other elements which are located on the perimeter of the array are chosen to be parasitic.

Figure 6 shows the results of the fully active array elements and the proposed array with only four active elements. For the former array, with all its elements being active, the D was 17.1 dBi, SLL was −14.4 dBi, and HPBW was 25.5°, while for the latter array with only four active elements, the D was 11.2 dBi, the SLL was −28.5 dBi, and the HPBW was 49.8°.

In the next example, an array with 5 × 5 elements is considered. For the proposed array, the elements that were located on the inner rectangular ring are chosen to be active (i.e., a number of eight elements), while the remaining elements are chosen to be parasitic as shown in **Figure 7**. For fully active array elements, the D = 19.1 dBi, SLL = −13.9 dBi, and HPBW = 20.3°, while the values for the proposed array are found to be D = 14 dBi, SLL = −13.9 dBi, and HPBW = 31.5°.

In all of the aforementioned examples, although the gain of the proposed array is relatively smaller than that of the fully active array elements, the radiation patterns are still in an acceptable shape. Moreover, these radiation patterns are produced with a very simple feeding network. Furthermore, the peak of the first sidelobe was significantly reduced down compared to that of the fully active array elements.

Finally, the performance of the proposed array with four active elements on the center of an array with size 6 × 6 elements is considered. The performance in terms

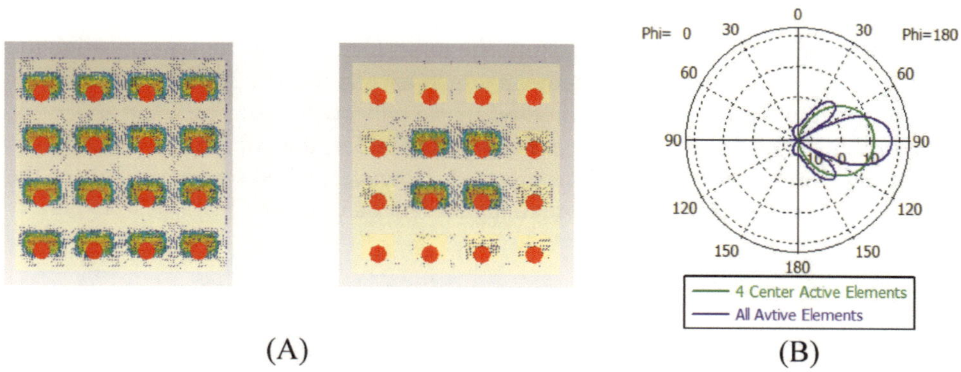

Figure 6.
Results for 4 × 4 planar arrays: (A) current distributions on both active and parasitic elements and (B) radiation patterns in the polar plot.

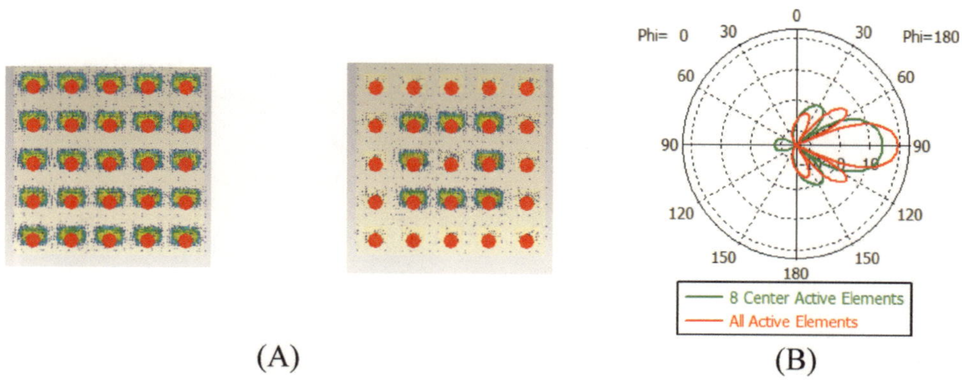

Figure 7.
Results for 5 × 5 planar arrays: (A) current distributions on both active and parasitic elements and (B) radiation patterns in the polar plot.

Spacing in (mm)	Spacing in (λ)	Directivity (dBi)	SLL (dBi)	HPBW (°)
43.75	0.35 × λ	10.60	−26.05	54.5
56.25	0.45 × λ	11.21	−35.80	50.9
62.50	0.50 × λ	11.30	−37.60	50.4
68.75	0.55 × λ	11.30	−33.95	50.6
81.25	0.65 × λ	11.32	−30.86	47.8
93.75	0.75 × λ	11.60	−34.20	46.6

Table 2.
The performance of the proposed array versus separation distances between active and parasitic elements.

of D, SLL, and HPBW is investigated under various values of the separation distances between the active and the parasitic elements as shown in **Table 2**.

5. Conclusions

Various planar array configurations are designed to produce radiation patterns with various characteristics, such as various beamwidth and low sidelobe level. It is shown that the radiation pattern of a small number of the active elements can be improved by properly selecting the most efficient configuration of the parasitic elements which are usually chosen from the outer elements of the array. Unlike the existing array pattern synthesis methods which usually use the digital phase shifters and/or attenuators, the proposed method needs only switches to turn the array elements either on (active) or off (parasitic). Therefore, it is robust to the excitation errors and simpler and cheaper to implement. However, when some elements of the array are switched off, then the obtained gain will be reduced.

Author details

Jafar Ramadhan Mohammed* and Karam Mudhafar Younus
College of Electronics Engineering, Ninevah University, Mosul, Iraq

*Address all correspondence to: jafarram@yahoo.com

References

[1] Balanis CA. Antenna Theory, Analysis and Design. 4th ed. Hoboken, New Jersey: Wiley; 2016

[2] Mailloux RJ. Phased Array Antenna Handbook. 2nd ed. Norwood, MA: Artech House; 2005

[3] Haupt RL. Antenna Arrays: A Computational Approach. Hoboken, New Jersey: John Wiley and Sons; 2010

[4] Mohammed JR. Thinning a subset of selected elements for null steering using binary genetic algorithm. Progress in Electromagnetics Research. 2018;**67**: 147-157

[5] Ismail TH, Dawoud MM. Null steering in phased arrays by controlling the element positions. IEEE Transactions on Antennas and Propagation. 1991;**39**:1561-1566

[6] Mohammed JR. Optimal null steering method in uniformly excited equally spaced linear arrays by optimizing two edge elements. Electronics Letters. 2017; **53**(13):835-837

[7] Mohammed JR. Element selection for optimized multi-wide nulls in almost uniformly excited arrays. IEEE Antennas and Wireless Propagation Letters. 2018;**17**(4):629-632

[8] Rodriguez-Gonzalez JA, Ares-Pena F. Design of planar arrays composed by an active dipole above a ground plane with parasitic elements. Progress in Electromagnetics Research. 2011;**119**: 265-277

[9] Hemant P, Gautam KM. Design of non-uniformly spaced circular arrays of parasitic dipoles for lower side lobe level with maximum directivity. Advanced Electromagnetics. 2018;**7**(1):51-56

Chapter 6

New Radiation Pattern-Reconfigurable 60-GHz Antenna for 5G Communications

Yasir I.A. Al-Yasir, Hasanain A.H. Al-Behadili, Baha A. Sawadi, Naser Ojaroudi Parchin, Ahmed M. Abdulkhaleq, Abdulkareem S. Abdullah and Raed A. Abd-Alhameed

Abstract

Reconfigurable beam steering using circular disc microstrip patch antenna with a ring slot is proposed. The overall dimension of the antenna is 5.4 × 5.4 mm^2 printed on 0.504 mm thick, RT5870 substrate with relative permittivity 2.3 and loss tangent 0.0012. The designed antenna operates at the expected 60 GHz 5G frequency band with a central coaxial probe feed. Two NMOS switches are utilized to generate three different beam patterns. Activating each switch individually results in a 70° shift in the main beam direction with constant frequency characteristics. The power gain is 3.9–4.8 dB in the three states of switch configurations. Simulated results in terms of return loss, peak gains and radiation pattern are presented and show good performance at the expected 60 GHz band for 5G applications.

Keywords: beam steering, 5G antennas, microstrip antennas, NMOS, reconfigurable antennas

1. Introduction

Ideally, reconfigurable antennas must be able to change their operating frequency, polarization and radiation pattern independently to provide varying operating requirements [1]. Moreover, the progress of these antennas faces significant challenges to both antenna modelling and system designer. However, these challenges will come not only from the antenna design only but also from the surrounding technologies that enable reconfigurability [2–5].

Nowadays, there exist extensive research areas covering reconfigurable antennas usually that concerned on the design and the technological perspectives. Radiation pattern and polarization-reconfigurable antennas are also very attractive topics because they can provide various characteristics which lead to a better signal to noise ratio (SNR) as well as higher quality of service (QoS) of the whole systems.

Recently, theoretical and measurement results of 5G communications operating at mmWave were studied and discussed [6, 7]. In [7], the researchers presented and designed the unprecedented hardware challenges and essential design considerations concerned with the antenna system methodology for future communications. The designed antenna provides a good and useful isotropic radiated power with minimum loses for practical use in 5G cellular communications.

In modern wireless communications systems, antennas may be required to have several angles of the main beams, and one of the methods to achieve this property is using multiple elements of a single antenna in array forms to make pattern diversity [6]. Pattern diversity of an array can be accomplished by setting the so-called array factor [7]. However, mutual coupling between antenna elements in array antennas has a significant effect on wireless radio links, causing undesirable effects on system performance. Thus, one of the efficient methods to overcoming these drawbacks is using radiation pattern-reconfigurable antennas. These antennas can diverse the main lobe direction or null direction at specific angles. In addition, it provides a wider coverage area with less interference and noise signals by controlling the main beam locations.

Generally, PIN diodes [8–10] and RF-MEMS [10, 11] are used as switching elements in order to alter the antenna radiation patterns. In [9], pattern and frequency reconfigurable annular slot antenna were presented. Matching stubs are used to achieve a planar antenna operating with three different frequencies, 5.2, 6.4 and 5.8 GHz. PIN diodes are used as switches to activate or deactivate the stubs, creating a reconfigurable matching network. In [12], the author explained the integration of the radio frequency microelectromechanical system (RF MEMS) switches with radiation pattern-reconfigurable antennas. In another design [13], a radiation pattern switchable parasitic array antenna at 60 GHz band was proposed. NMOS transistor switches are used to alter the main beam direction of the antenna with a constant frequency. Beam scanning angle of this antenna can deviate at $\pm 56^{\circ}$.

However, the radiation pattern-reconfigurable antenna must be able to alter its radiation pattern without a significant change in other characteristics such as operating frequency, impedance bandwidths, and polarizations, which can be considered as the first challenge facing the designers of the radiation pattern-reconfigurable antenna [1]. There are another two challenges for the simulation tool necessary to modelling the reconfigurable antennas [14]:

1. Modelling the switch requires a new design technology considering all the switch characteristics. Where the practical switches may have different properties and factors to be considered to get more agreement between the simulation and measurement performance.

2. The same antenna with multiple operating configurations, all of them might affect the performance of the others, must be designed and optimized simultaneously. Thus, a single structure has to be modelled to perform all different state of operations, and this involves a professional simulation technology.

The main difference between the present article, and that of other researchers dealing with reconfigurable planar antennas, is that a new design technique is used to simulate, modelling and optimize the structures, leading to a reconfigurable radiation pattern with constant impedance bandwidth and polarization characteristics at mmWave spectrum.

The work presented here introduces a new pattern-reconfigurable planar circular disk microstrip antenna capable of controlling the main beam direction using two NMOS switches. The design investigations initiate from a conventional disk antenna with a ring slot. The operational configurations are achieved through independently controllable switches, in which, each one of them is implemented as an

NMOS switch. The designed antenna alters the radiation pattern according to three switch combinations, which can alternate the main beam into three different radiation angles. The proposed antenna is printed on the top of a 0.504 mm thickness of Roger RT5870 substrate with relative permittivity 2.3 and loss tangent 0.0012 of size 5.4 × 5.4 mm^2. The radiation pattern of the designed antenna alters about 70° in yz-plane at 60 GHz frequency. CST studio simulation software [15] is used to optimize antenna and simulate the design, with hybrid co-simulation between CST MWS and CST DS used to implement the designed antenna taking into account the SPICE model for the NMOS switch and the effect of the biasing circuit.

2. Radiation pattern-reconfigurable antenna application

The range of wireless communication technologies that have emerged in the last two decades is immense. Nowadays, the tendency is to study new methods of integration and interoperability between devices. The electronics for wireless devices are being designed in the perspective of shrinking and group multiple functions in one single integrated circuit; also, the software and firmware on the processing units are being developed to be faster and less energy consuming. Besides streamlining the electronics and processing techniques, one can do improvements in the radio interface, namely in the antennas [16].

There are already different solutions to address the interoperability capabilities of the antennas. From multi-band antennas to wideband and ultra-wideband antennas, many approaches have been presented in the past years. However, this type of antennas has some disadvantages. Multi-band antennas are not versatile, they are difficult to design and not always present a good performance for all the bands, besides, and most of them are not very small, which can arouse size concerns. Ultra-wideband antennas can cover very large frequency bandwidths with seeming performance, however, this performance is not good for any particular band, which leads to low moderate gains and they are quite large as well [17].

Another way to achieve versatility is with the application of reconfigurable antennas, which allow reducing the number of antennas present in a given device, ensuring the interoperability between systems. However, this can also be integrated with more complex systems such as MIMO (multiple-input multiple-output) and/or cognitive radio systems, also known as SDR (Software Defined Radio) [18, 19]. The concept of antenna selection in MIMO systems may be explored in terms of reconfigurable antenna elements as shown in **Figure 1**. The complexity in terms of feed chains is less for the reconfigurable setup with a reduced number of antenna elements. The different configurations of the reconfigurable antennas give the antenna array a more possible option than with a conventional two-element array.

Reconfigurability allows an antenna to adapt in real time different parameters like the resonant frequency, the polarization or even the radiation pattern to some extent. These are extremely useful features for mobile devices when considering the expansion of the different communication systems in the near future. Given the versatility and advantages these antennas can provide for new wireless devices, these have been subject of much attention in the investigation field in the past few years and there are already many different solutions presented to acquire different reconfigurability to different systems. Reconfigurable antennas are designed to be implemented on various platforms which cover various wireless services that are spanned over a wide frequency range. In particular, reconfigurable antennas are proposed for higher efficiencies in various implementations that involve cognition and continuous adaptation to the environment such as in cognitive radio and MIMO systems.

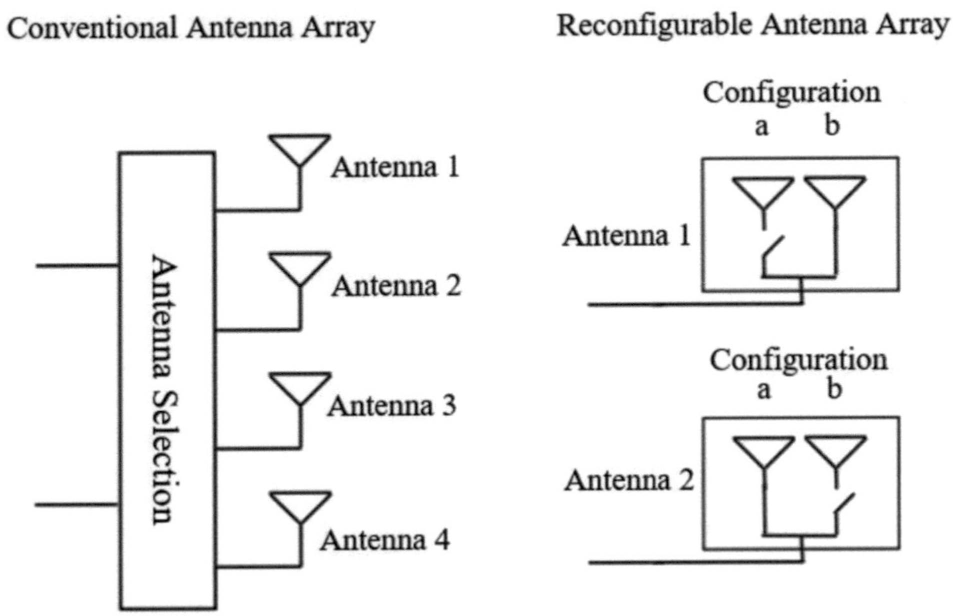

Figure 1.
Block diagram representing a MIMO antenna array with conventional antenna elements (left) and reconfigurable antenna elements (right).

A MIMO system employs multiple antennas at both the transmitter and the receiver frontends. The advantage of using such configurations is that different information can be sent simultaneously, thereby increasing the communication spectral efficiency in a multipath environment. According to the varying channel conditions and user's need, a MIMO system can adjust the modulation level, coding rate, and the transmission signalling schemes. Radiation pattern/polarization reconfigurable antennas add an additional degree of freedom in a MIMO environment and thus improve the system performance. The use of this type of antennas increases the capacity significantly by allowing the selection between different pattern diversity and polarization configurations. Reconfigurable antennas can also be used in modern space applications. In such cases, it is required to be able to reconfigure the antenna radiation pattern to serve a new coverage zone, limit fading in rainy areas, and maintain high data rate at as many frequency bands as possible [20].

Most of the advanced wireless communications' applications (cognitive radio, MIMO, Space communications) require highly efficient software controlled dynamic antennas. These antennas that can be reconfigured using software and based on users' requests have to be highly reliable as well as able to achieve the required functions. Therefore, an antenna designer proposing a new reconfigurable antenna design for an advanced application has to present a design that has minimal losses in its structure and operation. The antenna has to exhibit software accessibility as well as be able to be controlled using programmable controllers. Most importantly, a designer must ensure that the proposed antenna is able to operate under unforeseen circumstances continuously; thus, a reliability study is required from reconfigurable antenna designers.

3. Radiation pattern-reconfigurable antenna design

The first step is to design the antenna without the inclusion of reconfiguring property. This achieved by designing a circular patch antenna, by adopting

Rogers RT5870 as the dielectric substrate with h = 0.504 mm, $\varepsilon_r = 2.3$, tangent loss = 0.0012, and $L_{sub} = W_{sub} = 5.4$ mm. The antenna is designed and optimized to operate at a particular frequency of 60 GHz. This frequency is chosen because it is suitable and widely used in 5G applications. Other than the rectangular patch, the configuration used in this paper is the circular patch or disk, as shown in **Figure 2**, where R represents the actual disk radius. Based on the cavity model formulation [9], a design procedure is outlined which leads to practical designs of circular microstrip antennas for the dominant TM^z_{110} mode.

The procedure is as follows:

The actual radius of the patch (R) at the resonant frequency f_0 can be calculated as [9]:

$$f_0 = \frac{1.8412 \times C}{2\pi R\sqrt{\varepsilon_r}} \tag{1}$$

where *c* is the speed of light in free-space?

The resonant frequency of Eq. (1) does not take into account the fringing effect. Therefore, for the circular patch, a correction is introduced by using an effective radius R_e, to replace the actual radius R as:

$$R_e = R\left\{1 + \frac{2h}{\pi\varepsilon_r h}\left[ln\left(\frac{\pi R}{2h}\right) + 1.7726\right]\right\}^{1/2} \tag{2}$$

Thus, the resonant frequency of Eq. (1) should be modified by using Eq. (2) and expressed as:

$$f_0 = \frac{1.8412 \times C}{2\pi R_e\sqrt{\varepsilon_r}} \tag{3}$$

Then, a first-order approximation to the solution of Eq. (2) for R is given by:

$$R = \frac{F}{\left\{1 + \frac{2h}{\pi\varepsilon_r F}\left[ln\left(\frac{\pi F}{2h}\right) + 1.7726\right]\right\}^{1/2}} \tag{4}$$

where,

$$F = \frac{8.791 \times 10^9}{f_0\sqrt{\varepsilon_r}} \tag{5}$$

Figure 3 shows the schematic diagram of the proposed antenna. An annular slot ring circular disk planer antenna is designed. The feeding network used here

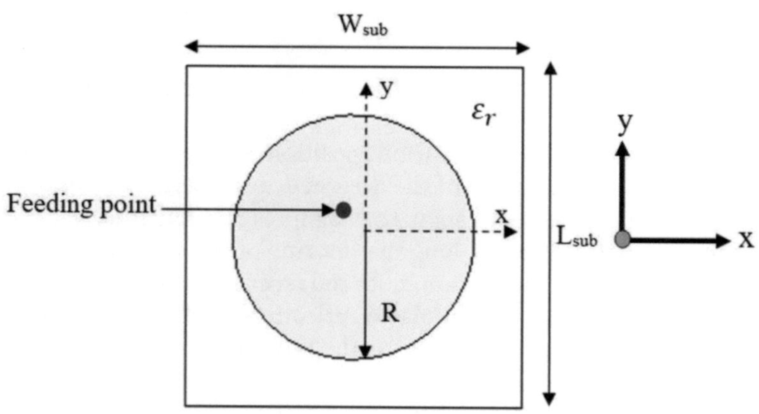

Figure 2.
Circular patch antenna.

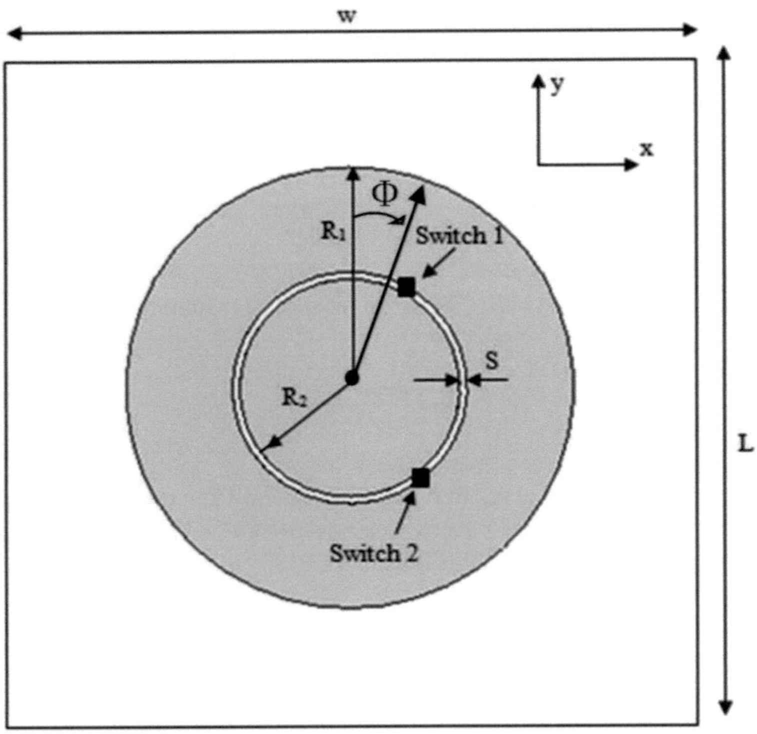

Figure 3.
The geometry structure of the proposed antenna.

W (mm)	L (mm)	S (mm)	R_1 (mm)	R_2 (mm)
5.4	5.4	0.06	1.85	0.91

Table 1.
Detailed dimensions of designed antenna shown in ***Figure** 3.*

is a coaxial probe with a radius equal to 0.07 mm. The circular disk has an outer radius R_1 = 1.85 mm and the inner radius R_2 = 0.91 mm. Detailed dimensions of the proposed antenna are shown in **Table 1**. The dimensions of this antenna are optimized to operate at the resonance frequency 60 GHz to cover 2 GHz bandwidth with a return loss less than 10 dB.

4. Parametric study and a hybrid EM-circuit co-simulation

In this section, the influences of various positions of the switches on both the response and radiation pattern of the designed antennas are discussed. Firstly, the switches are modelled as a metal tab that is represented by the ideal state of the switches. The metal tab is rotated along the slot ring of the antenna by an angle φ (φ = 00, 300, 1500, 1800). In addition, both the return loss and radiation pattern are observed. **Figure 4** shows the simulated reflection coefficient as a function of frequency for various sweep angles of φ. As the position of the switch is shifted, the resonance frequency of the antenna changes slightly around 60 GHz. In fact, this frequency shifting can be neglected since the resonance bandwidth around 60 GHz is kept unaffected. On the other hand, the effect of the parameter sweep on the radiation pattern in the yz-plane is studied and presented in **Figure 5**. It is clear that the reconfigurable property is achieved at two important and symmetrical angles (φ = 30 and 150°) along yz-plane.

Secondly, two NMOS transistor are used as switches to replacing the metal tab, in which, switch 1 is located at a position shifted by (φ = 30°) from the y-axis, whenever the switch 2 is shifted by (φ = 150°) from the y-axis. It should be noted that by using the computer simulation technology (CST), the NMOS transistor switches are modelled with a lumped element network with R_{on} = 10.2 Ω, C_{on} = 47.1 pF, R_{off} = 13.3 MΩ, C_{off} = 39 pF as shown in **Figure 6**. The figure shows the equivalent circuit for the switches in the ON and OFF states.

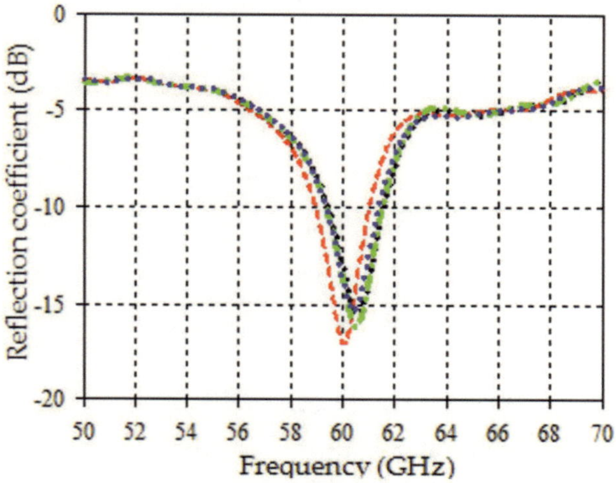

Figure 4.
The frequency response of the input reflection coefficient for different angles of φ using the metal tab as a switch (φ = 0: dotted line, φ =30: solid line, φ = 150: dashed line and φ = 180: dotted/dashed line).

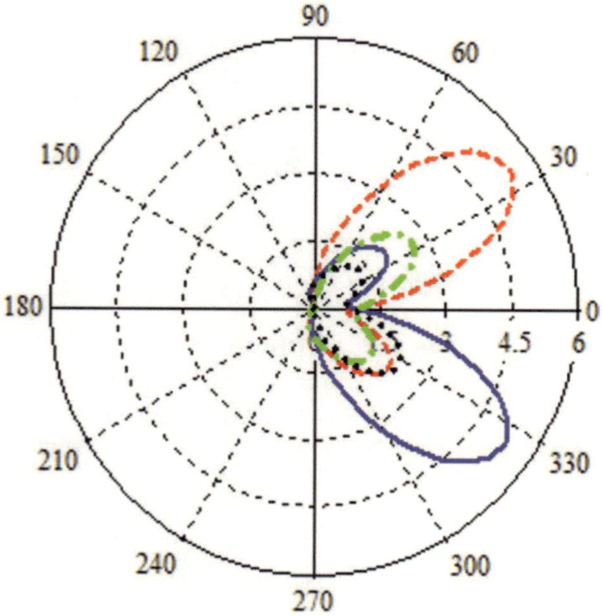

Figure 5.
Radiation pattern at yz-plane for different angles of φ using the metal tab as a switch (φ = 0: dotted line, φ =30: solid line, φ = 150: dashed line and φ = 180: dotted/dashed line).

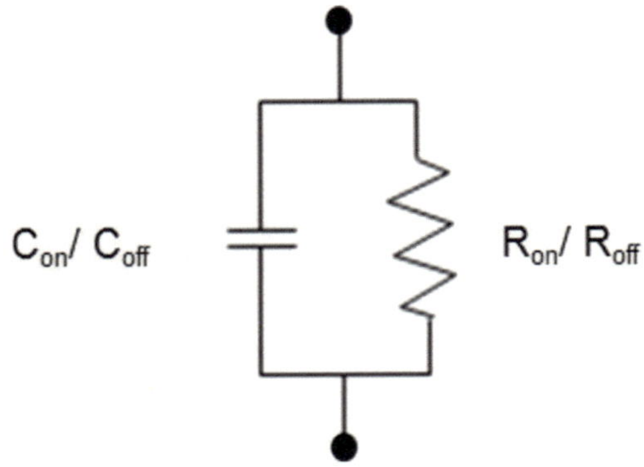

Figure 6.
Simplified electrical model of the NMOS transistor switch in ON and OFF states.

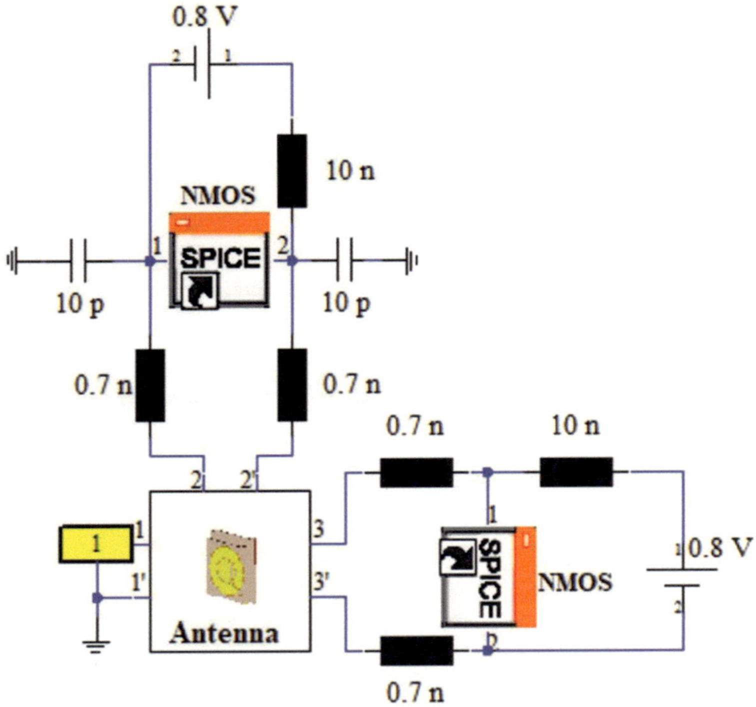

Figure 7.
A hybrid EM-circuit co-simulation of the antenna.

Finally, the co-simulation between CST MWS and CST DS for the designed radiation pattern-reconfigurable antenna are modelled and achieved, as shown in **Figure 7**. The simulation takes into account the spice model for the NMOS switches. The manufacturer-specified parasitic inductance and capacitance of the packaging are included as well.

5. Simulation results

The simulation results achieved by the CST microwave studio software. The performances of the proposed antenna, in terms of return losses, radiation patterns, efficiencies and gains, with different states of switches have been studied as follows:

5.1 Impedance bandwidth for S11 < −10 dB

Figure 8 shows the simulated results for the designed antenna. It is clear that (SW1 ON, SW2 OFF) state has −18 dB reflection coefficients at resonance frequency 60 GHz with impedance bandwidth of 3.4%. In (SW1 OFF, SW2 ON) state, the reflection coefficient is −16 dB at 61 GHz resonance frequency, with corresponding impedance bandwidth of 3.3%. Whenever, in (SW1 ON, SW2 ON) state, S_{11} is equal to −30 dB at 60.8 GHz resonance frequency, with impedance bandwidth 5.7%. The main difficulty of the designing of radiation pattern-reconfigurable antenna is that this type of reconfigurability must be accomplished without significant changes in the impedance or frequency characteristics. The achieved results overcome this difficulty by maintaining a single resonance frequency for all states of configurations.

5.2 Radiation patterns and directivities

Figures 9–11 show the simulated radiation patterns at 60 GHz in the yz-plane (E-plane). When the proposed antenna operates at (SW1 ON, SW2 OFF) state, the beam's maximal direction in the yz-plane is 35°. At (SW1 OFF, SW2 ON) state, the beam's maximal direction in the yz-plane is -35°. At (SW1 ON, SW2 ON) state, the beam's maximal direction in the yz-plane in ($35, -35^{\circ}$). ("+" indicates that the radiation pattern tilts toward the positive y-axis and "-" indicates that the radiation pattern tilts toward the negative y-axis).

As a result, the radiation patterns of the proposed antenna operating at different switching states can divers by 70° shifted along yz-plane (E-Plane). Due to the symmetry characteristics on both the antenna structure and switches positions, the single main beam direction can be altered symmetrically around the z-axis in the yz-plane as shown in **Figures 9–11**. **Figures 12–14** show the 3D simulation results for beam directivities, which correspond to the three configurations of the switches in both ON and OFF states.

5.3 Antenna gains and efficiencies

Simulation results for the realized gains of the proposed antenna as a function of frequency are shown in **Figure 15**. Simulated maximum realized gain at (SW1 ON, SW2

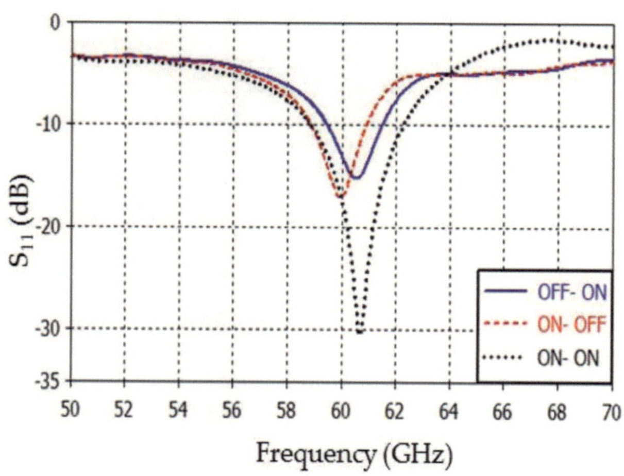

Figure 8.
Simulation results for the proposed antenna for different states of switching.

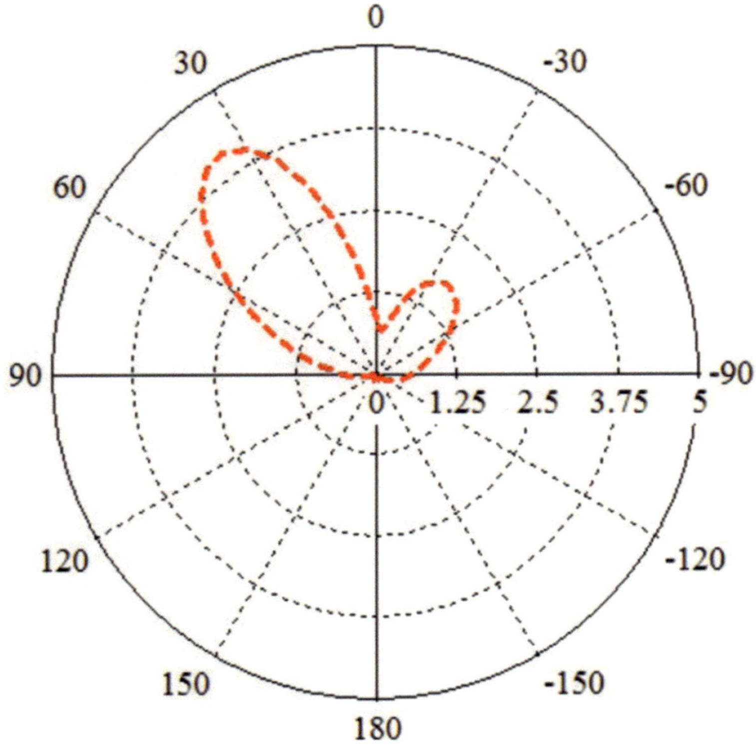

Figure 9.
Radiation pattern at yz-plane in the ON-OFF state.

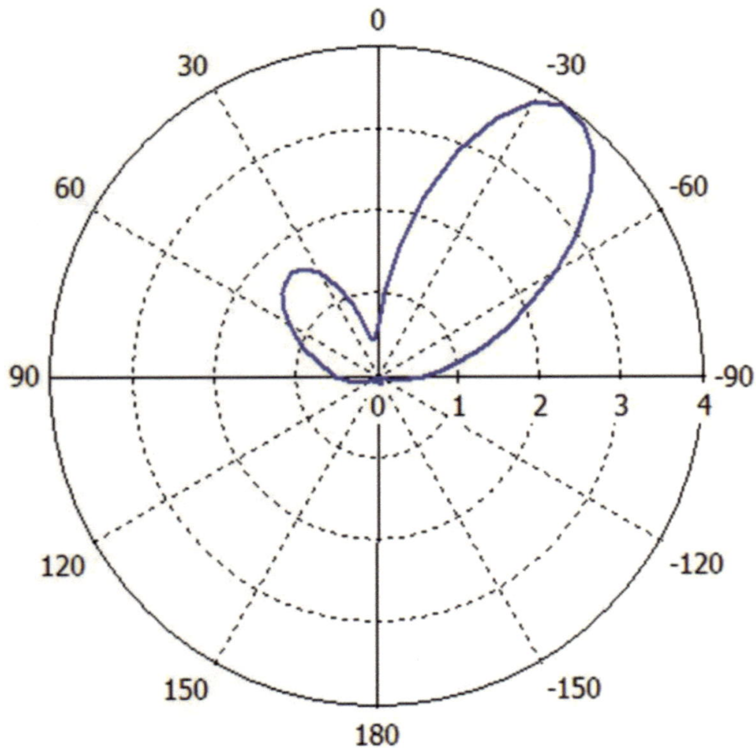

Figure 10.
The radiation pattern at yz-plane in the OFF-ON state.

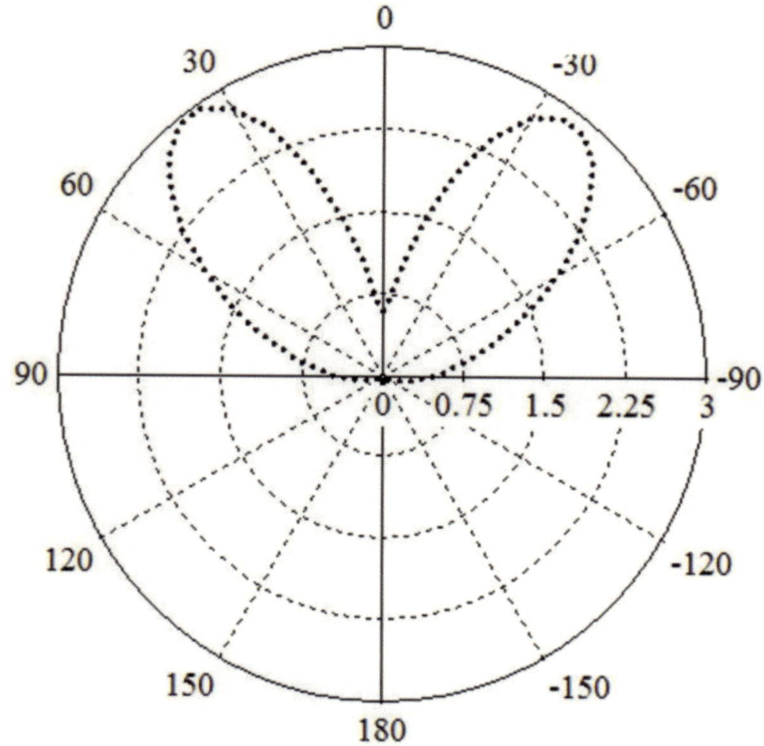

Figure 11.
Radiation pattern at yz-plane in the ON-ON state.

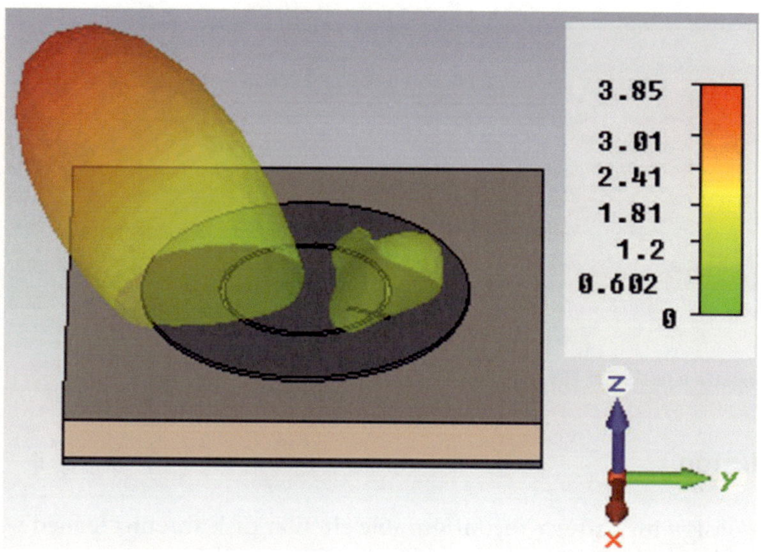

Figure 12.
3D radiation pattern in the ON-OFF state.

OFF) is 4.5 dB. At (SW1 OFF, SW2 ON) state, the maximum gain is 4.8 dB. In addition, at (SW1 ON, SW2 ON) state, the maximum gain and directivity are 3.9 dB. Total efficiency for the designed antenna is studied and plotted as a function of frequency in the useful bandwidth of 60 GHz at different state of each switch as shown in **Figure 16**. It is clear that the achieved efficiencies are better than 95% at all the states of switches.

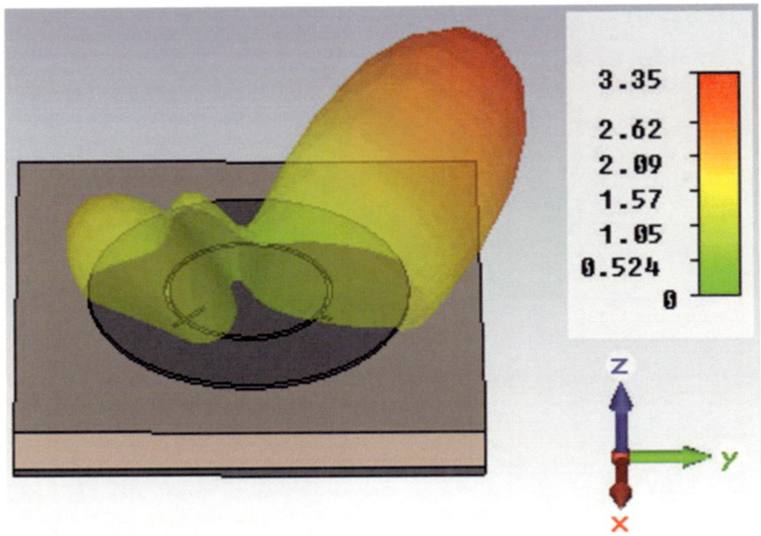

Figure 13.
3D radiation pattern in the OFF-ON state.

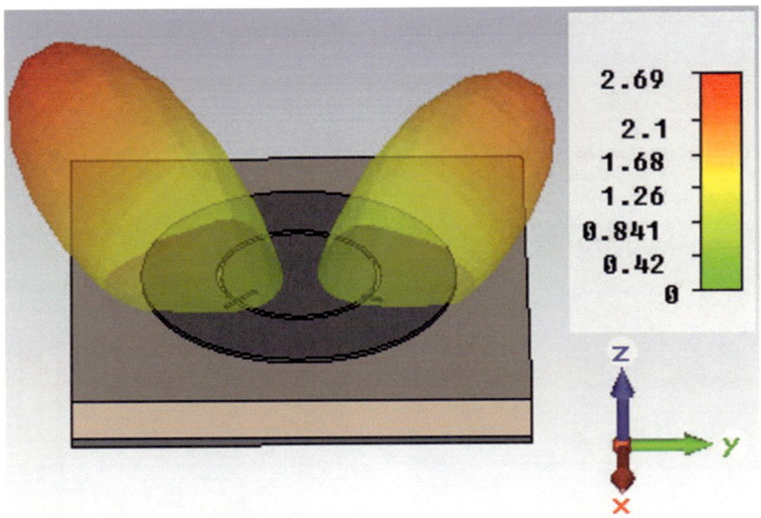

Figure 14.
3D radiation pattern in the ON-ON state.

6. Conclusion

A new design for pattern-reconfigurable circular disk antenna loaded with an annular slot ring has been presented for 60 GHz mmWave applications. The antenna is reconfigurable between three different radiation patterns by employing two NMOS switches. The achieved results show that the proposed antenna can redirect the main beam at −35 and 35° in the yz- plane with the maximum achieved gains are 4.5, 4.8 and 3.9 dB, and impedance bandwidths about 3.4, 3.3 and 5.7% in the (SW1 ON, SW2 OFF), (SW1 OFF, SW2 ON) and (SW1 ON, SW2 ON) states, respectively, at the resonance frequency 60 GHz. Antenna's efficiency better than 95% was achieved in the three states of switching configurations. The antenna beam pattern characteristics, efficiencies, peak gains and impedance bandwidths are

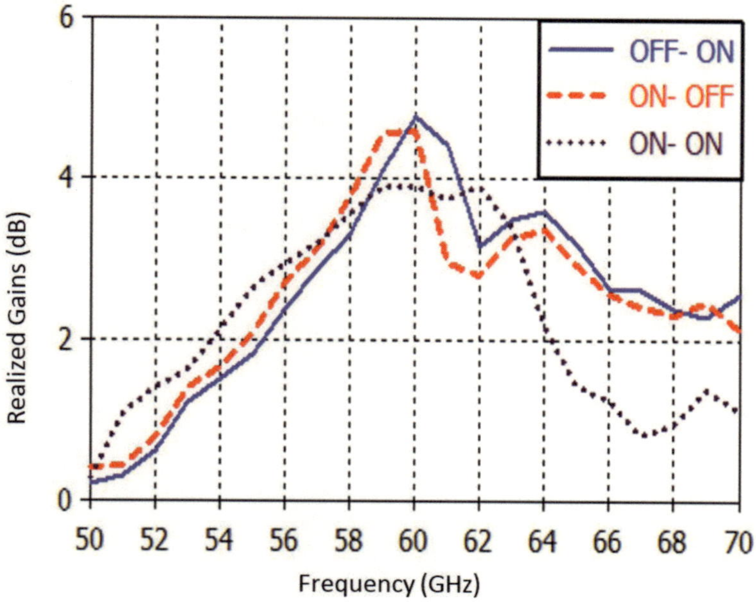

Figure 15.
Realized gain in different states of switches for the proposed antenna.

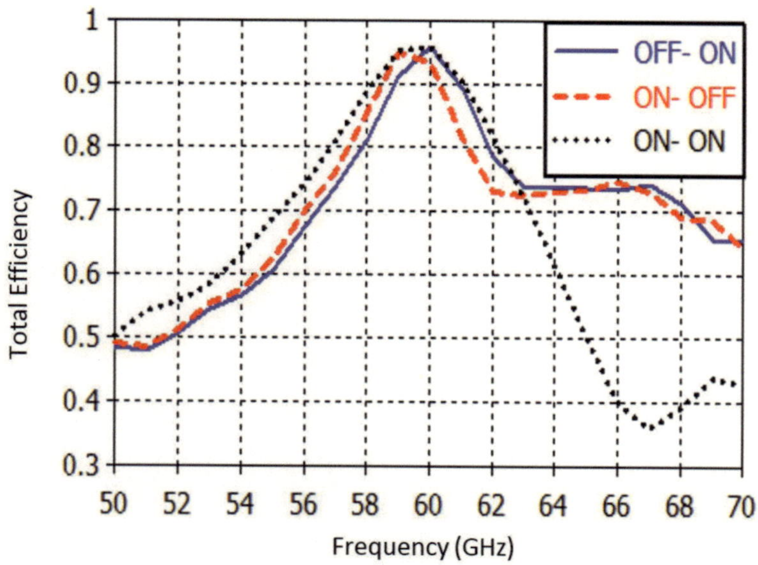

Figure 16.
Total efficiency in different states of switches for the proposed antenna.

suitable for 5G applications. Also, due to the simple construction and beam pattern diversity, the designed antenna can find different applications in MIMO systems.

Acknowledgements

This project has received funding from the European Union's Horizon 2020 research and innovation programme under Grant agreement H2020-MSCA-ITN-2016 SECRET-722424.

Author details

Yasir I.A. Al-Yasir[1*], Hasanain A.H. Al-Behadili[2], Baha A. Sawadi[3], Naser Ojaroudi Parchin[1], Ahmed M. Abdulkhaleq[1], Abdulkareem S. Abdullah[4] and Raed A. Abd-Alhameed[1]

1 Faculty of Engineering and Informatics, University of Bradford, Bradford, UK

2 Department of Electrical Engineering, Faculty of Engineering, University of Misan, Misan, Iraq

3 Department of Communication Engineering, Iraq University College, Basra, Iraq

4 Department of Electrical Engineering, College of Engineering, University of Basra, Basra, Iraq

*Address all correspondence to: y.i.a.al-yasir@bradford.ac.uk

References

[1] Bernhard JT. Reconfigurable Antennas. San Rafael, CA: Morgan and Claypool; 2007

[2] Haider N, Caratelli D, Yarovoy AG. Recent developments in reconfigurable and multiband antenna technology. International Journal of Antennas and Propagation. 2013;**2013**:14

[3] Christodoulou GC, Tawk Y, Youssef A, Lane AS, Scott RS. Reconfigurable antennas for wireless and space applications. Proceedings of the IEEE. 2012;**100**(7):2250-2261

[4] Abdulraheem YI et al. Design of frequency reconfigurable multiband compact antenna using two PIN diodes for WLAN/WiMAX applications. IET Microwaves, Antennas and Propagation. 2017;**11**(8):1098-1105

[5] Al-Yasir YIA, Abdullah AS, Ojaroudi Parchin N, Abd-Alhameed RA, Noras JM. A new polarization-reconfigurable antenna for 5G applications. Electronics. 2018;7:293

[6] Roh W et al. Millimeter-wave beamforming as an enabling technology for 5G cellular communications: Theoretical feasibility and prototype results. IEEE Communications Magazine. 2014;**52**(2):106-113

[7] Hong W, Baek K-H, Lee Y, Kim Y. Study and prototyping of practically large-scale mm wave antenna systems for 5G cellular devices. IEEE Communications Magazine. 2014;**52**(9)

[8] Skrivervik AK, Mosig JR. Finite phased array of microstrip patch antennas: The infinite array approach. IEEE Transactions on Antennas and Propagation. 1992;**40**(05):579-582

[9] Balanis CA. Antenna Theory. 3rd ed. USA: A John Wiley and Sons, Inc.; 2005. pp. 283-369

[10] Kahn WS, Park JA, Yoon YJ. Simple reconfigurable antenna with radiation pattern. Electronics Letters. 2008;**44**(3):182-183

[11] Nikolaou S, Bairavasubramanian R, Lugo C, Carrasquillo I, Thompson DC, Ponchak GE, et al. Pattern and frequency reconfigurable annular slot antenna using pin diodes. IEEE Transactions on Antennas and Propagation. 2006;**54**(2):439-448

[12] Huff GH, Bernhard JT. Integration of packaged RF-MEMS switch with radiation pattern reconfigurable square spiral microstrip antennas. IEEE Transactions on Antennas and Propagation. 2006;**54**(2):464-469

[13] Nair SSV, Dussopt L, Siligaris A. Design of a reconfigurable 60-GHz on-chip CMOS-SOI pattern-diversity antenna. In: European Conference on Antennas and Propagation (EuCAP). 2013. pp. 2825-2828

[14] Rutschlin M, Sokol V. Reconfigurable antenna simulation: Design of reconfigurable antennas with electromagnetic simulation. IEEE Microwave Magazine. 2013;**14**(7):92-101

[15] Microwave Studio Based on the Finite Integration Technique. Framingham, MA, USA: CST; 2014

[16] Chang K. RF and Microwave Wireless Systems. New York: Wiley; 2000

[17] Aberle JT, Oh SH, Aukland DT, Roegers SD. Reconfigurable antenna and RF front-ends for portable wireless devices. IEEE Antennas and Propagation Magazine. 2003;**45**(6):148-154

[18] Tsoulos GV. MIMO System Technology for Wireless Communications. Boca Raton: CRC Press; 2006

[19] Fette B. Cognitive Radio Technology. Elsevier: Newnes; 2006

[20] Rao S, Minh T, Chien Hsu C. Reconfigurable antenna system for satellite communications. In: IEEE International Symposium on Antennas and Propagation. Honolulu, Hawaii. June 2007. pp. 3157-3160

Chapter 7

Ultra Wide Band Antenna with Defected Ground Plane and Microstrip Line Fed for Wi-Fi/Wi-Max/DCS/5G/Satellite Communications

Ashish Singh, Krishnananda Shet and Durga Prasad

Abstract

In this chapter, ultra wide band angular ring antenna has been proposed for wireless applications. It has been observed that antenna resonate from 2.9 to 13.1 GHz which has 10.2 GHz bandwidth. Further, it is observed that antenna has nearly omni-directional radiation pattern for E and H-plane at 3.5, 5.8, and 8.5 GHz. The theoretical analysis of the proposed has been done using circuit theory analysis. It was also found using simulation that antenna has good input and output response of 0.2 ns. Proposed antenna measured, simulated, and theoretical results matches for antenna characteristics, i.e., reflection coefficient and radiation pattern. Bandwidth of antenna lies between 2.9 and 13.1 GHz, so this antenna is suitable for Wi-Fi, Wi-Max, digital communication system (DCS), satellite communication, and 5G applications.

Keywords: ultra wide band (UWB), angular ring, finite ground plane, microstrip line fed, circuit theory

1. Introduction

Wireless communication systems are highly desired in various fields of security systems, Wi-Fi, Wi-Max, and mobile communication. These applications have special common devices, i.e., an antenna for efficient transmission and reception information. Presently, antennas are equipped in all communication devices and these devices are size and volume constrains. This leads to reduction of size for antenna in existing communication device. All communication devices have patch antennas for transmission and reception of signals. Scientists and researchers are investigating on these antennas since 1972 for reduction in size and increase in bandwidth. To achieve this, numbers of patch antenna designs and techniques were proposed. It was found by researchers that angular ring patch antenna is an efficient antenna which gives both size reduction and increased bandwidth.

Angular ring patch antenna was first reported in year 1985, by IJ Bahal for biomedical application. Thereafter, only few research have been reported on these antennas such as, theory and experiment on the annular-ring microstrip antenna, shared aperture microstrip patch antenna array for L and S-Bands, analysis of a gap-coupled stacked annular-ring microstrip antenna, compact stacked circularly polarized annular-ring microstrip antenna for GPS applications, annular-ring microstrip patch antenna with finite ground plane for ultra-wideband applications, compact concentric annular-ring patch antenna for triple-frequency operation, comparison of several novel annular-ring microstrip patch antennas for circular polarization, analysis of two-concentric annular-ring microstrip antenna, and broadband circularly polarized annular-ring microstrip antenna [1–10]. All above reported papers has some limitations such as complicated geometry, lacks theoretical analysis for defected ground with microstrip line feed, circuit diagram at Radio frequency were not proposed for designed antenna and theoretical, simulated and experimental results were not compared.

In this chapter, ultra-wideband microstrip patch antenna is proposed for Wi-Fi, Wi-Max, DCS, and 5G applications. Partial ground plane with microstrip line fed techniques is used to achieve UWB band for various wireless applications. Detail descriptions of proposed antenna design are discussed in next section.

2. Geometrical consideration

The microstrip line fed angular ring patch antenna with rectangular ground is shown in **Figure 1** and the antenna is fabricated on FR_4 substrate and its top and bottom view is shown in **Figure 2**. The proposed antenna has been designed on FR_4 substrate of height "h" and overall dimension of designed geometry is $(12 \times 14 \times 1.57)$ mm^3. The proposed strip line fed angular ring antenna comprises of ground plane of dimension $(W_G \times L_G)$, and strip line of dimension $(W_L \times L_L)$. Further, antenna is excited via SMA connector fed via strip line. The design specification of complete antenna design is given in **Table 1**. Fabricated antenna picture

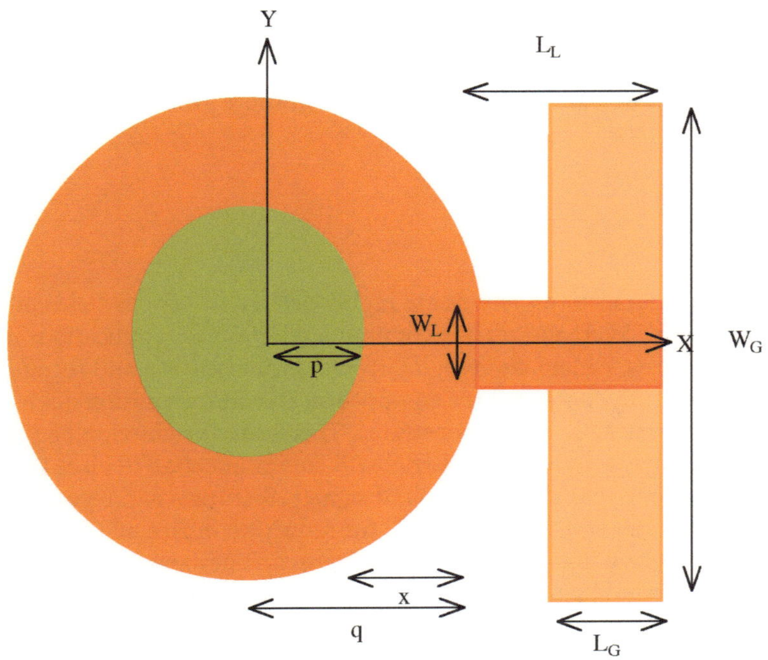

Figure 1.
Radiating structure for proposed antenna.

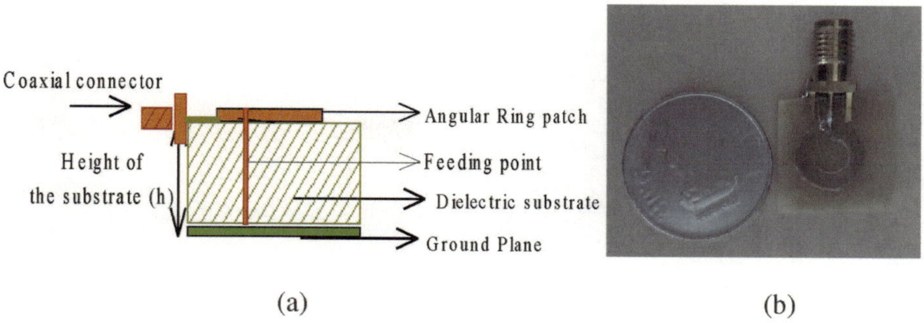

Figure 2.
Designed antenna (a) top view (b) fabricated antenna on FR4.

FR-4 lossy, ε_r	4.4
Radius of angular ring inner, p	3.5 mm
Radius of angular ring outer, q	8.5 mm
Circular path, x	5.0 mm
Strip line length, L_L	5.0 mm
Strip line width, W_L	3.5 mm
Ground plane width, W_G	2.5 mm
Ground plane length, L_G	12 mm
Height of the substrate, h	1.57 mm

Table 1.
Design specification of angular ring antenna.

is shown in **Figure 2(b)**. It can be observed from figure that antenna is very compact in size and can be utilized for compact communication devices.

3. Theoretical investigations

The resonating frequency of angular ring [11] is given as

$$f = \frac{\chi_{nm}}{2\pi p\sqrt{\epsilon_r}}, \tag{1}$$

where c is the velocity of light in free space, $\chi_{nm} = k_{nm}p$, k_{nm} is for the resonant TM_{nm} modes.

The inner and outer radii of angular ring are given as $p_e = p - (x_e - x)/2$ and $q_e = q - (x_e - x)/2$, respectively. The p_e, q_e, and x_e are the effective increase in length of inner, outer, and path width, respectively.

The angular ring patch antenna can be represented in circuit diagram as combination of inductance, capacitance and conductance, as shown in **Figure 3**. The values of inductance L, capacitance C, and conductance G are calculated as.

$$L = \frac{\mu h}{\pi k^2[n,m]}\left[J_n(k_1e)G'_n(k_1p_e) - G_n(k_1e)J'_n(k_1p_e)\right], \tag{2}$$

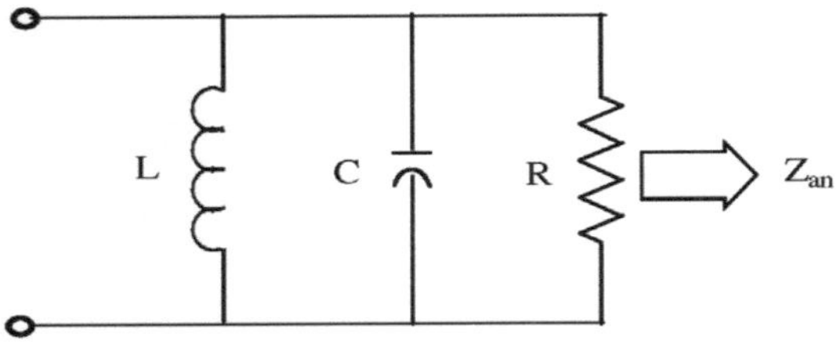

Figure 3.
RF circuit representation of angular ring.

$$C = \frac{\mu \epsilon_0 \epsilon_r}{L k_1^2}, \tag{3}$$

$$G = \frac{1}{R} = Re\left[\frac{\pi}{h}\left\{\left(\frac{E_p}{E_z}\right)^2 g(p,p) + \left(\frac{E_q}{E_c}\right)^2 g(q,q) - \frac{2E_p E_q}{E_z} y(p,q)\right\}\right], \tag{4}$$

where $[n,m] = \frac{1}{2k_1^2}\left[(k_1^2 q_e - 1)\ \{J_n(k_1 q_e)Y'_n(k_1 p_e) - Y_n(k_1 q_e)J'_n(k_1 p_e)\} - \frac{4}{\pi^2 k_1^2 p_e}(k_1^2 q_e^2\text{-}1)\right]$, $y(p,q)$ is mutual admittance for angular ring between inner and outer radii, $g(p,p)$ is conductance across inner periphery of angular ring, $g(q,q)$ is conductance across outer periphery of angular ring, E_p is the radiation field around inner periphery of angular ring, E_q is the radiation field at outer periphery of angular ring, E_z is the radiation field considered due to ground plane.

Input impedance for the angular ring is given as,

$$Z_{an} = \frac{1}{G + j\omega C + 1/j\omega L}, \tag{5}$$

Angular ring is connected to strip line; the strip line can be represented into RF circuit as combination of L_s, C_s, and Z_0. The RF circuit of the strip line angular patch antenna is shown in **Figure 4**, where L_s and C_s are inductance and capacitance of strip [12, 13].

$$L_s = 100h\left(4\sqrt{W_S/h} - 4.21\right), \text{nH} \tag{6}$$

$$C_s = W_s\{(9.5\varepsilon_r + 1.25)W_s/h + 5.2\varepsilon_r + 7.0\}, \text{pF} \tag{7}$$

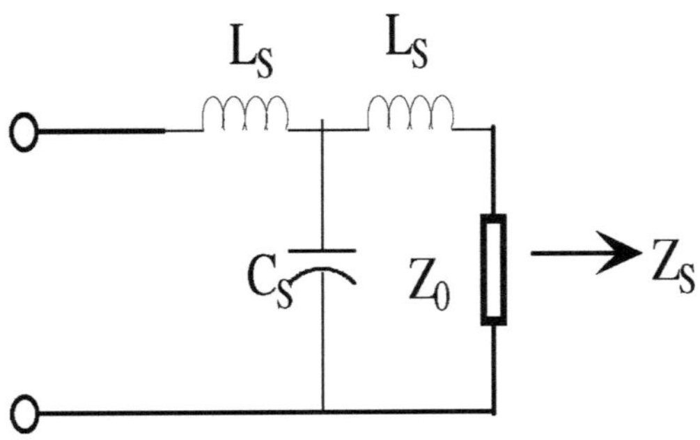

Figure 4.
RF circuit diagram for strip line.

DOI: http://dx.doi.org/10.5772/intechopen.91428

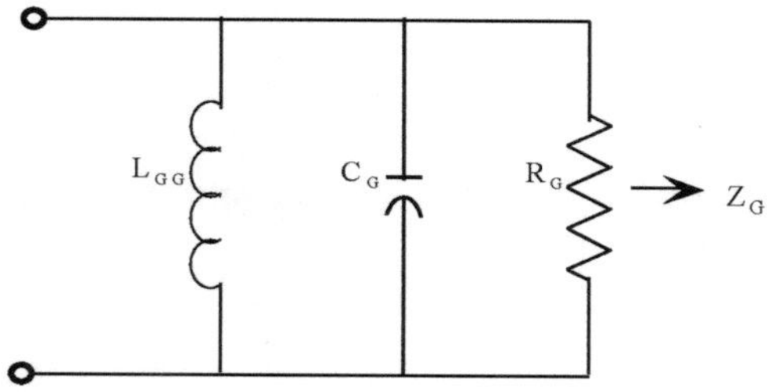

Figure 5.
RF circuit representation for ground plane.

The resonating frequency of the strip line is given as

$$f = c/2L_{se}\sqrt{\varepsilon_{re}}, \tag{8}$$

The ground plane patch is represented as RF circuit combination of Resistance R_G, inductance L_{GG}, and capacitance C_G. The RF circuit representation of the ground plane is shown in **Figure 5**, R_G, C_G, L_{GG} can be calculated as [12, 13].

$$C_G = \frac{L_G W_G \varepsilon_0 \varepsilon_e}{2L_G}\cos^2\left(\frac{\pi}{L_G}\right), \tag{9}$$

$$R_G = \frac{Q}{\omega_r^2 C_G}, \tag{10}$$

$$L_{GG} = \frac{1}{C_G \omega_r^2}, \tag{11}$$

$$\text{Quality factor}, Q = \frac{c\sqrt{\varepsilon_e}}{4fh}. \tag{12}$$

ε_e is the effective permittivity of the medium.

Further, there will be strong electromagnetic field between top patch (i.e., angular ring and strip line) and ground plane (rectangular patch). Due to which inductance and capacitance are developed between them and its RF circuit representation is shown in **Figure 6**. Thereafter, on excitation of antenna the impedance is also developed between top and bottom patch [12–14] and represented as,

$$Z_{ee} = 1/\left[\frac{1}{j\omega L_{ee}} + j\omega C_{ee}\right] \tag{13}$$

$$L_{ee} = \frac{L_g \times L_{an}}{L_g + L_{an}} + L_{gan}, \tag{14}$$

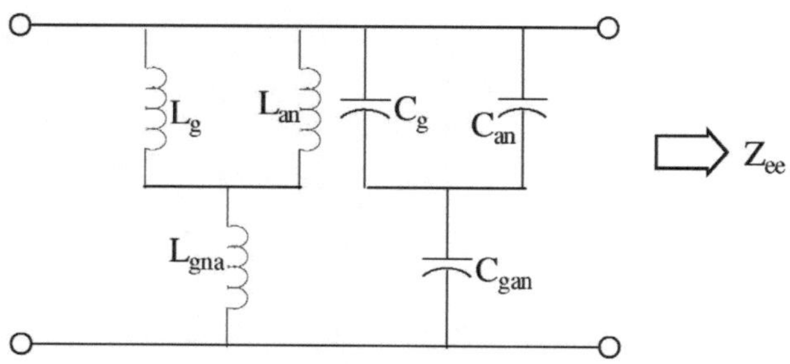

Figure 6.
RF circuit diagram of electromagnetically coupled between ground plane and radiating patch.

$$C_{ee} = \frac{(C_g + C_{an}) \times C_{gan}}{C_g + C_{an} + C_{gan}} \tag{15}$$

L_{ee} and C_{ee} are the equivalent inductance and capacitance developed because of electromagnetic coupling.

L_{an} and C_{an} are electromagnetically developed mutual inductance and capacitance for angular ring.

$$\mathrm{L_{gan}} = \frac{\mathrm{k}_{\mathrm{gc}}^2 (L_g + L_{an}) + \left[k_{gc}^4 (L_g + L_{an})^2 + 4k_{gc}^4 \left(1 - k_{gc}^2\right) L_g L_{an}\right]^{1/2}}{2\left(1 - k_{gc}^2\right)}, \tag{16}$$

$$C_{gan} = \frac{-(C_g + C_{an}) + \left[(C_{an} + C_g)^2 + \left(1 - 1/k_{gc}^2\right) C_{an} C_g\right]^{1/2}}{2}, \tag{17}$$

$$k_{gc} = \frac{1}{\sqrt{Q_g Q_{gg}}}, \tag{18}$$

$$Q_g = R_G \sqrt{\frac{C_G}{L_{GG}}}, \tag{19}$$

$Q_{gg} = R_{an} \sqrt{\frac{C_{an}}{L_{an}}}$, Q_g and Q_{gg} are quality factor for both the resonators, $\mathrm{R_{an}}$ impedance of microstrip.

The input impedance of strip line feed angular ring with defected ground plane and its RF circuit representation is shown in **Figure** 7 and is calculated using Eqs. (2)–(19)

$$Z_{in} = \frac{1}{\frac{1}{Z_G} + \frac{1}{Z_s} + \frac{1}{Z_{an}} + \frac{1}{Z_{ee}}} \tag{20}$$

Using Eq. (20) the input impedance of proposed has been used to calculate reflection coefficient (RC), return loss (RL) and voltage standing wave ratio (VSWR) can be calculated as,

$$\text{Reflection Coefficient } \Gamma = \frac{Z - Z_{in}}{Z + Z_{in}}, \tag{21}$$

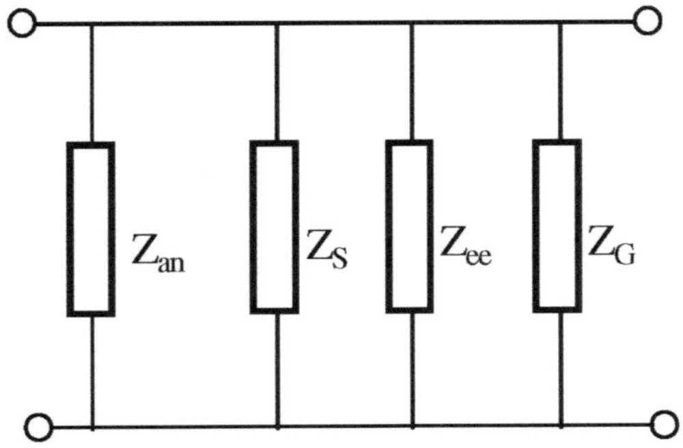

Figure 7.
The equivalent RF circuit diagram of microstrip angular patch antenna.

where Z is the input impedance of the microstrip fed (*50 Ω*).

$$VSWR = \frac{1 + |\Gamma|}{1 - |\Gamma|}, \tag{22}$$

$$\text{Return loss (RL)} = 20 \log |\Gamma| \tag{23}$$

Using Eqs. (21)–(23) the theoretical plots for RL, VSWR, and RC with respect to frequency (GHz) can be plotted. In this chapter, theoretical plot for RC with respect to frequency (GHz) is plotted.

4. Result and discussion

The comparison between simulated [14], measured, and theoretical results is shown in **Figure 8(a)**. It is observed from the figure that these three results are in approximately close with each other. Further, the bandwidth of theoretical, measured and simulated results is 10.63, 10.6, and 10.2 GHz, respectively. Bandwidth of antenna lies between 2.9 and 13.1 GHz, these bands are suitable for Wi-Max, Wi-Fi,

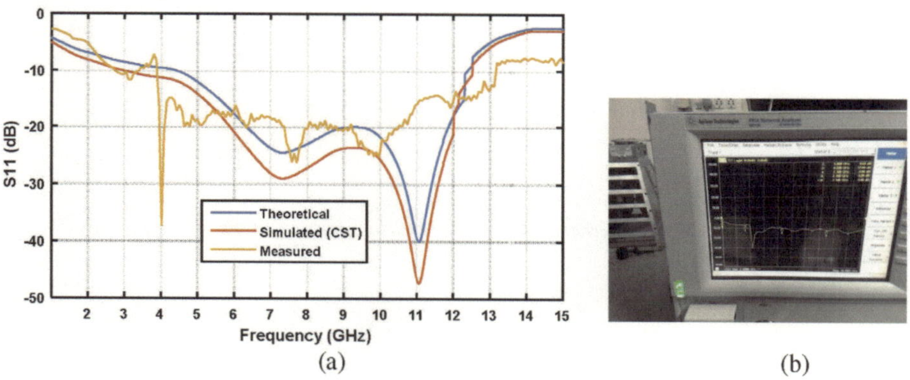

Figure 8.
(a) Comparative results of proposed antenna; (b) measured result on VNA.

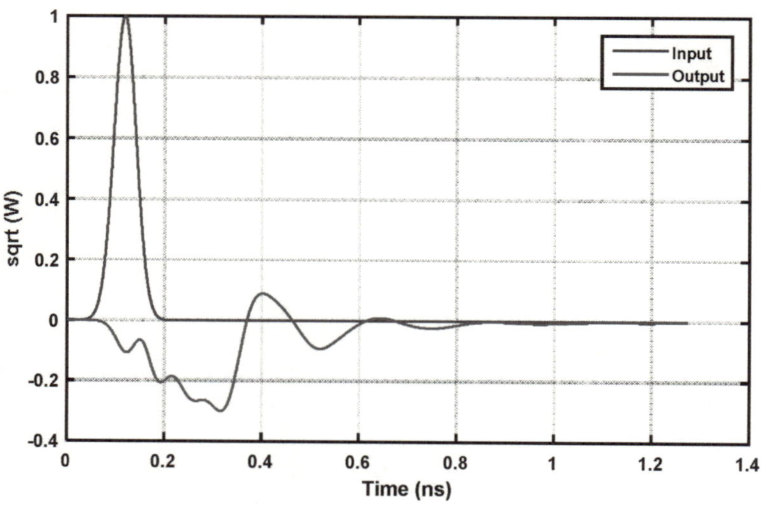

Figure 9.
Input and output response for the excited proposed antenna.

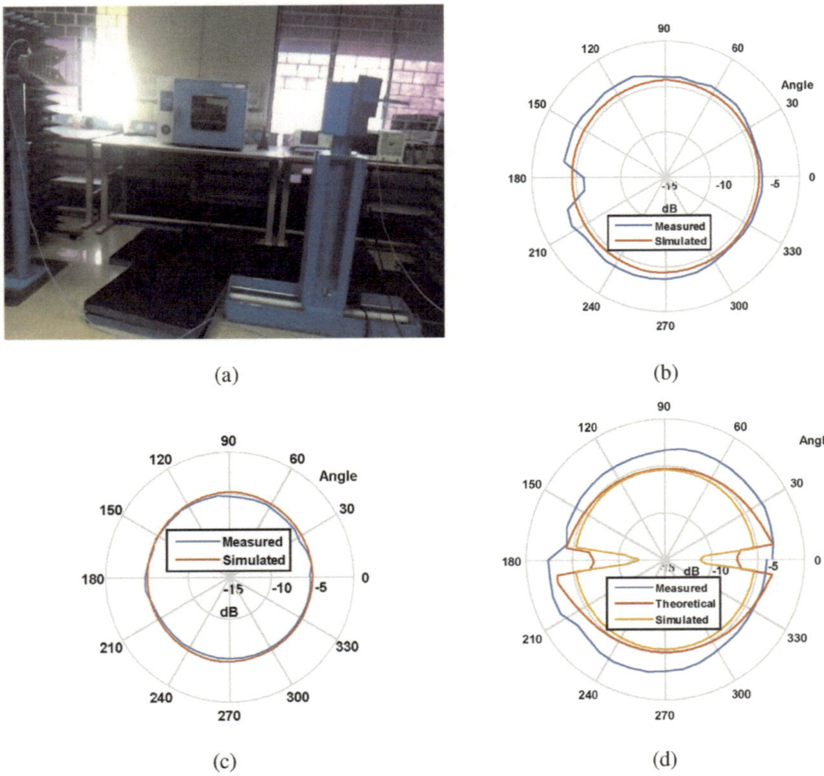

Figure 10.
Radiation pattern for H-plane (a) anechoic chambers with antennas; (b) 3.5 GHz, (c) 5.8 GHz, and (d) 8.5 GHz.

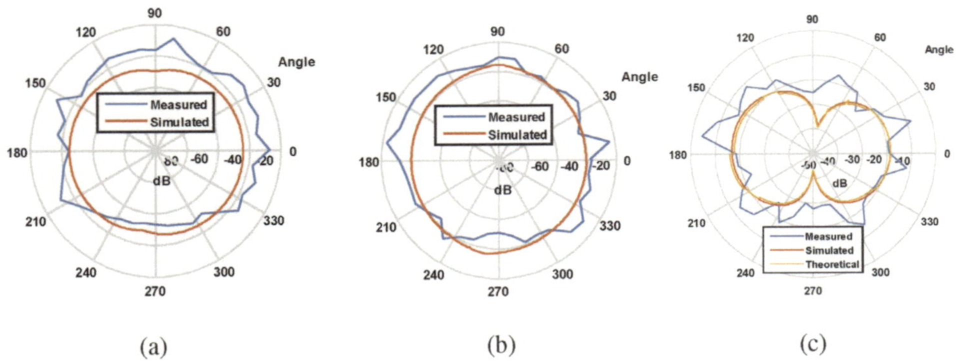

Figure 11.
Radiation pattern for E-plane (a) 3.5 GHz, (b) 5.8 GHz, and (c) 8.5 GHz.

digital communication system, satellite communication, and 5G applications. Measured result picture on Vector Network Analyzer (VNA) of proposed antenna is shown in **Figure 8(b)**.

From **Figure 9**, it is observed that maximum input amplitude of the antenna is 1 sqrt W at 0.1 ns time interval, whereas maximum output amplitude is 0.34 sqrt W at a response time of 0.38 ns with phase reversal. The output response is not smooth because angular ring structure with rectangular ground plane.

Radiation pattern of the proposed antenna are shown in **Figures 10** and **11** for E and H-plane, respectively. **Figure 10(a)** shows the radiation pattern measurement setup, proposed antenna under test is kept 200 cm apart from the horn antenna. **Figure 10(b)** and **(c)** are measured and simulated radiation pattern at 3.5 and 5.8 GHz, respectively. Measured and simulated radiation pattern are in close agreement in both cases and omni-directional patterns are observed. **Figure 10(d)** shows radiation pattern at center frequency. Major and minor lobes have been observed of same beam width for measured, theoretical, and simulated antenna at center frequency 8.5 GHz for H-plane. **Figure 11(a)–(c)** shows the radiation pattern for E-plane at 3.5, 5.8, and 8.5 GHz, respectively. Antenna shows nearly omni-directional radiation pattern at 3.5 and 5.8 GHz; whereas at 8.5 GHz, it is partially eight shaped. Electric field intensity is more toward 180° for E-plane and antenna 3 dB beam width is 87.4°. Slightly mismatch is observed in radiation pattern results because of partially open anechoic chambers and fabrication defects. It has omni-directional pattern so this antenna can be utilized for mobile communication.

Figures 12–14 show the surface current distribution of proposed antenna at center frequency for 90 and 0° phase, respectively. Further, from **Figure 12**, the

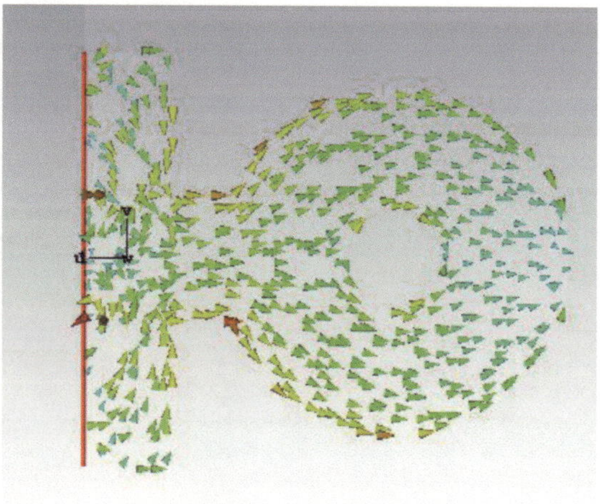

Figure 12.
Current distribution at 90° phase.

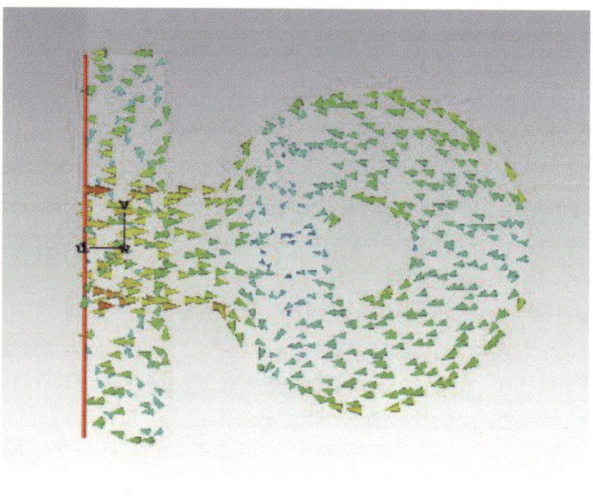

Figure 13.
Current distributions at center frequency at 0° phase.

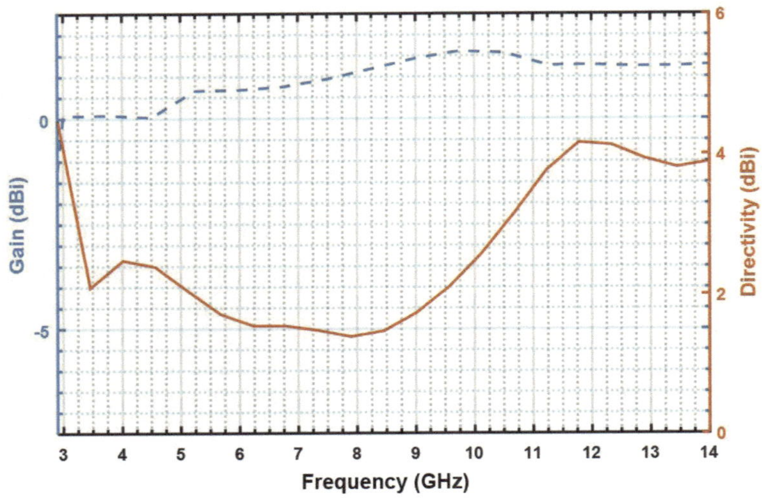

Figure 14.
Gain and directivity of proposed antenna.

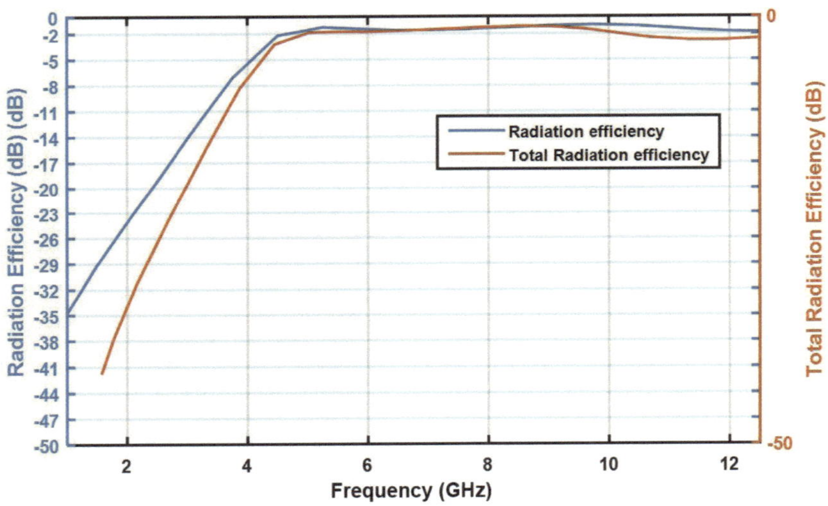

Figure 15.
Radiation efficiency of proposed antenna.

maximum surface current of 86.4766 A/m is observed at the outer perimeter of angular ring near the edges of strip line and surface current is evenly distributed along length of antenna at 90° phase. Whereas, the surface current at 0° phase is not evenly distributed along the length of antenna and more surface current observed near feed as shown in **Figure 13**.

Figure 14 shows the gain and directivity of the antenna in dBi. It is observed that maximum gain of 2.75 dBi is achieved at 10.2 GHz; whereas, average gain of antenna is 2.1 dBi. Further, the directivity of the antenna at 10.2 GHz is maximum, i.e., 4.1 dBi and average directivity is 2.98 dBi.

The maximum radiation efficiency is achieved −1.5 dB (70.79%) at 10.2 GHz frequency as observed in **Figure 15** and total radiation efficiency of proposed antenna at 10.2 GHz is found to be (−2.2 dB) 60.25%. This is because the loss occurs due to skin effect and conduction loss in antenna device.

5. Conclusion

From the above theoretical analysis, it is found that angular ring patch antenna can be utilized for UWB antenna. Further was observed that antenna cover frequency band between 2.9 and 13.1 GHz which has 125% bandwidth. From the results, it was also observed that antenna has good radiation characteristics and input and output response. Antenna has the gain and efficiency of 2.2 dBi and 70.79%. Simulated, measured, and theoretical results are matching for radiation pattern and reflection coefficient. Further, this antenna is suitable for digital communication system, satellite communication, and 5G applications.

Acknowledgements

The authors would like to thanks Nitte Education Trust for providing the research Grant No. Res/NMAMIT/03. We would like to thank Dr. K. Krishnamoorthy, Department of Electronics and Communications, National Institute of Technology, Surathkala, for providing measurement faculties of antenna in their research lab.

Author details

Ashish Singh*, Krishnananda Shet and Durga Prasad
Department of Electronics and Communication Engineering, N.M.A.M. Institute of Technology (Affiliated to Visvesvaraya Technological University, Belagavi) Nitte, Udupi, Karnataka, India

*Address all correspondence to: ashsin09@rediffmail.com

References

[1] Lee KF, Dahele JS. Theory and experiment on the annular-ring microstrip antenna. Annales des Telecommunications. 1985;**40**:508-515

[2] Dhiman J, Sharma A, Khah SK. Shared aperture microstrip patch antenna array for L and S-bands. Progress in Electromagnetics Research Letters. 2019;**86**:91-95

[3] Ansari JA, Ram RB, Singh P. Analysis of a gap-coupled stacked annular ring microstrip antenna. Progress in Electromagnetics Research B. 2008;**4**: 147-158

[4] Ding K, Yu T-B, Zhang Q. A compact stacked circularly polarized annular-ring microstrip antenna for GPS applications. Progress in Electromagnetics Research Letters. 2013;**40**:171-179

[5] Rawat S, Sharma KK. Annular ring microstrip patch antenna with finite ground plane for ultra-wideband applications. International Journal of Microwave and Wireless Technologies. 2015;7(2):179-184

[6] Bao XL, Ammann MJ. Compact concentric annular-ring patch antenna for triple-frequency operation. Electronics Letters. 2006;**42**:1129-1130

[7] Bao XL, Ammann MJ. Comparison of several novel annular-ring microstrip patch antennas for circular polarization. Journal of Electromagnetic Waves and Applications. 2006;**20**:1427-1438

[8] Kanaujia BK, Vishvakarma BR. Analysis of two-concentric annular ring microstrip antenna. Microwave and Optical Technology Letters. 2003;**36**: 104-108

[9] Guo Y-X, Bian L, Shi XQ. Broadband circularly polarized annular-ring microstrip antenna. IEEE Transactions on Antennas and Propagation. 2009;**57**: 2474-2477

[10] Rasool N, Kama H, Basit MA, Abdullah M. A low profile high gain ultra lightweight circularly polarized annular ring slot antenna for airborne and airship applications. IEEE Access. 2019;7:155048-155056

[11] Kumar G, Ray KP. Broadband Microstrip Antenna. USA: Artech House; 2003

[12] Bahal IJ, Bartia P. Microstrip Patch Antenna. USA: Artech House; 1980

[13] Meada M. Analysis of gap in microstrip transmission line. IEEE Transactions on Antennas and Propagation. 1972;**32**:1375-1379

[14] Computer Simulation Technology (CST). Microwave Studio Suite Version. Germany: Dassault Systèmes and Darmstadt; 2018

Chapter 8

Noise Characteristic Analysis of Multi-Port Network in Phased Array Radar

Yu Hongbiao

Abstract

Noise figure and noise power are detailedly analyzed and deduced in theory for multi-port network in active phased array radar. The mathematical expressions of output noise power and noise figure of network are given out under various conditions. Accordingly, this provides a basis of theories for multi-port network and radar receiver system design, the test method of array noise figure. Finally, two application examples are given to verify the accuracy of the formulae. Making use of these formulas, the designer can use to calculate the dynamic range of the radar receive system, and the designer can also constitute a measure scheme of the array noise figure for active phased array radar.

Keywords: noise power, noise figure, active network, passive network, active phased array radar

1. Introduction

In modern active phased array radar, the active antenna array is generally composed of dozens to tens of thousands of active transmit/receive (T/R) modules. However, the feeding of T/R modules (receiving echo signal and transmitting excitation power) is usually realized by a multi-port feeding network. The calculation of noise power and the measurement method of system noise figure of active antenna array including multi-port feed network are essential work for radar system designers and receiver designers. Understanding the analysis and calculation of noise power and noise figure of multi-port network is the basis for design specification such as system dynamic range, so how to correctly analyze and calculate noise power and noise figure of active antenna array is an important factor in radar system design. Next, the analysis and calculation of noise power for multi-port network and the calculation and measurement method of system noise figure in active phased array radar will be described in detail.

2. Noise power and noise figure of two-port network

For a two-port linear network as shown in **Figure 1**, suppose that $G(L)$ in the figure is the gain (loss) of the two-port network and NF is the noise figure of the two-port linear network. BW is the signal bandwidth, then the equivalent noise temperature of the two-port linear network is [1].

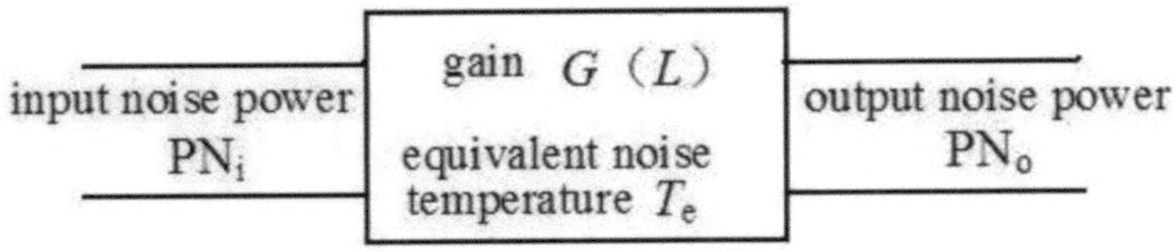

Figure 1.
Noise characteristics of a two-port network.

$$T_e = (\mathrm{NF} - 1)T_0 \tag{1}$$

where T_e is the equivalent noise temperature and T_0 is the room temperature, equal to 290 K.

The white noise power PN_0 is [2]:

$$\mathrm{PN}_0 = kT_0\mathrm{BW} \tag{2}$$

where k is the Boltzmann constant, equal to 1.381×10^{-23} J/K.

Assuming that the noise temperature of the input port is T_i, the input noise power PN_i is:

$$\mathrm{PN}_i = kT_i\mathrm{BW} \tag{3}$$

The noise power PN_a of a two-port linear network caused by equivalent noise temperature is:

$$\mathrm{PN}_a = kT_e\mathrm{BW} \tag{4}$$

The noise figure NF is expressed by noise power as:

$$\mathrm{NF} = 1 + \frac{\mathrm{PN}_a}{\mathrm{PN}_0} \tag{5}$$

The output noise power PN_o of the two-port linear network is:

$$\mathrm{PN}_o = G\mathrm{PN}_i + G\mathrm{PN}_a \tag{6}$$

Note that T_i in the formula is not necessarily equal to T_0.

3. Analysis of noise characteristics of passive two-port linear network

We first analyze the noise characteristics of the passive two-port linear network as shown in **Figure 2**. In the figure, L is the insertion loss of the passive two-port network, BW is the operating bandwidth of the passive two-port linear network, T_i is the noise temperature at the input of the passive two-port network, PN_i is the

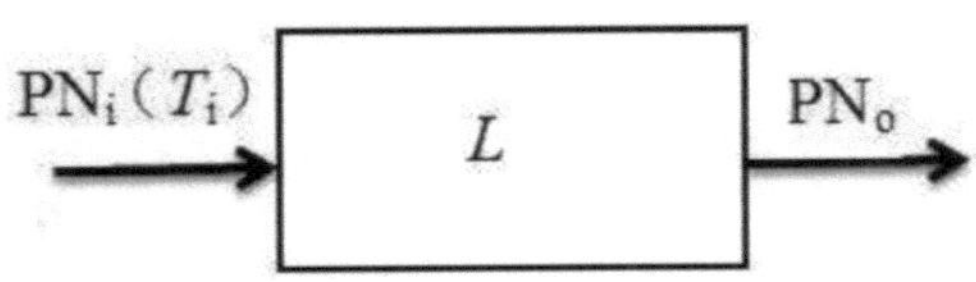

Figure 2.
Passive two-port linear network.

noise power at the input of the passive two-port network, and PN_o is the noise power at the output of the passive two-port network.

Let:

$$T_i = T_0 + T_{ei} \tag{7}$$

Then:

$$PN_i = kT_iBW = k(T_0 + T_{ei})BW \tag{8}$$

The noise power generated by PN_i at the network output is:

$$PN_{oi} = \frac{kT_iBW}{L} \tag{9}$$

The equivalent noise temperature of the passive two-port lossy network converted to the input of the two-port network is:

$$T_{eL} = (L-1)T_0 \tag{10}$$

The noise power generated by the passive two-port lossy network at the output is:

$$PN_{oL} = \left(1 - \frac{1}{L}\right) kT_0BW \tag{11}$$

Therefore, the total noise power generated by the passive two-port lossy network at the output is $PN_o = PN_{oi} + PN_{oL}$

$$\begin{aligned} PN_o &= \frac{kT_iBW}{L} + \left(1-\frac{1}{L}\right) kT_0BW = \frac{kT_0BW}{L} + \frac{kT_{ei}BW}{L} + \left(1-\frac{1}{L}\right) kT_0BW \\ &= kT_0BW + \frac{kT_{ei}BW}{L} \end{aligned} \tag{12}$$

4. Analysis of noise characteristics of multi-port linear passive networks

Next, we will analyze the noise characteristics of the multi-port linear passive network as shown in **Figure 3**. It is assumed that the multi-port linear passive network has n input ports and one output port, the active power loss of the network is L, and the signal bandwidth is BW.

Let the noise temperature of the jth input port of the multi-port linear passive network be:

$$T_{ij} = T_0 + T_{eij} \tag{13}$$

Then the noise power generated by the jth input port at the output of the multi-port linear passive network is:

$$PN_{oj} = \frac{k\left(T_0 + T_{eij}\right)BW}{nL} \tag{14}$$

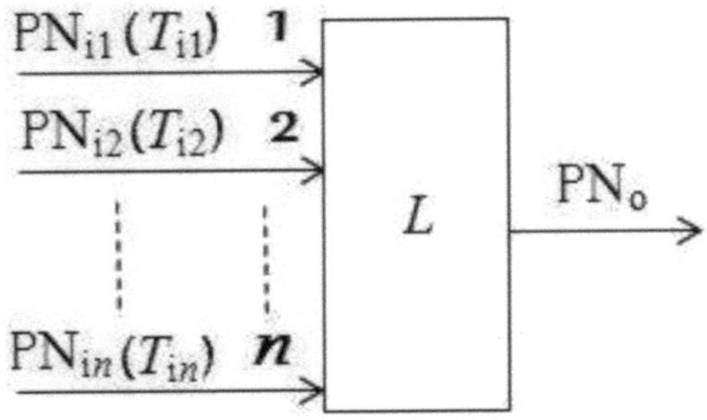

Figure 3.
Multi-port linear passive network.

Then the total noise power generated by the n input ports at the network output is:

$$\mathrm{PN}_{on} = \sum_{j=1}^{n} \mathrm{PN}_{oj} = \frac{k\mathrm{BW}}{nL} \sum_{j=1}^{n} \left(T_0 + T_{eij}\right) = \frac{kT_0\mathrm{BW}}{L} + \frac{k\mathrm{BW}}{nL} \sum_{j=1}^{n} T_{eij} \tag{15}$$

The equivalent noise temperature of the multi-port lossy passive network converted to each input is of the same formula (10), and then the noise power generated by the lossy network at the output is:

$$\mathrm{PN}_{oL} = \frac{nkT_{eL}\mathrm{BW}}{nL} = \left(1 - \frac{1}{L}\right) kT_0\mathrm{BW} \tag{16}$$

The total noise power generated at the output of the multi-port linear passive network is:

$$\mathrm{PN}_o = \mathrm{PN}_{on} + \mathrm{PN}_{oL} = kT_0\mathrm{BW} + \frac{k\mathrm{BW}}{nL} \sum_{j=1}^{n} T_{eij} \tag{17}$$

If T_{eij} of each input port of the multi-port linear passive network is the same as T_{ei}, then:

$$\mathrm{PN}_o = kT_0\mathrm{BW} + \frac{kT_{ei}\mathrm{BW}}{L} \tag{18}$$

By comparing Eq. (12) with Eq. (18), we can find that when T_{ei} = 0, i.e., each input port of the passive lossy network is connected to a matching load with a noise temperature of T_0, the noise power generated by the passive lossy network at the output port is equal, i.e., $\mathrm{PN}_o = kT_0\mathrm{BW}$.

5. Analysis of network noise characteristics after cascade of two-port active network and two-port passive network

If the two-port active network and the two-port passive network are cascaded, as shown in **Figure 4**, what is the noise characteristic of the cascaded two-port network? For the convenience of analysis, we make the noise figure of the two-port active network to be NF_1, the gain G, and the insertion loss L. For the convenience of analysis, it is assumed that the operating signal bandwidths of both are the same and both are BW.

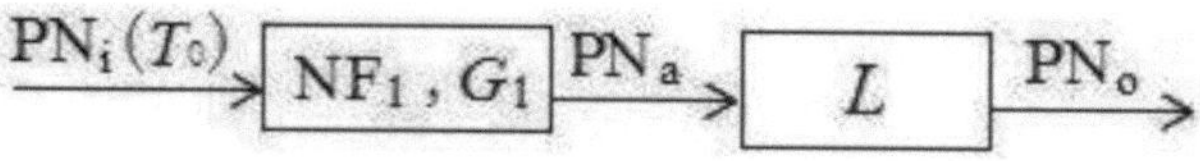

Figure 4.
Cascade of two-port active network and two-port passive network.

The equivalent noise temperature of the two-port active network converted to its input is [3]:

$$T_{\text{ea}} = (\text{NF}_1 - 1)T_0 \tag{19}$$

The equivalent noise temperature of a two-port passive network converted to its input is of the same formula (10). The noise power at the output of the two-port active network is thus:

$$\text{PN}_{\text{a}} = kT_0\text{BW}G + kT_{\text{ea}}\text{BW}G \tag{20}$$

The noise power PN_{a} at the output of the two-port active network produces the following noise power at the output of the passive network:

$$\text{PN}_{\text{oa}} = \frac{kT_0\text{BW}G}{L} + \frac{kT_{\text{ea}}\text{BW}G}{L} \tag{21}$$

The noise power generated by the two-port passive network itself at the output is shown in Eq. (11), so the total noise power at the output of the two-port synthetic network is:

$$\begin{aligned}\text{PN}_{\text{o}} = \text{PN}_{\text{oa}} + \text{PN}_{\text{oL}} &= \frac{kT_0\text{BW}G}{L} + \frac{kT_{\text{ea}}\text{BW}G}{L} + \left(1 - \frac{1}{L}\right)kT_0\text{BW} \\ &= kT_0\text{BW}\left(1 + \frac{G-1}{L}\right) + \frac{kT_{\text{ea}}\text{BW}G}{L}\end{aligned} \tag{22}$$

The total noise figure of the synthetic network after the two-port active network and the two-port passive network are cascaded is:

$$\text{NF} = \frac{\text{PN}_{\text{o}}}{kT_0\text{BW}G/L} = \frac{L + G - 1}{G} + \frac{T_{\text{ea}}}{T_0} = \text{NF}_1 + \frac{L-1}{G} \tag{23}$$

6. Analysis of network noise characteristics after cascaded two-port active network and multi-port passive network

Cascade n two-port active linear networks and a passive linear network with n input ports, as shown in **Figure 5**. Next, let's analyze the noise characteristics of n two-port active networks and multi-port passive networks after cascading.

As before, we assume that the active power loss of the multi-port passive network is L, the noise figure of the ith two-port active network is NF_{i}, and the gain is G_{i}. To facilitate analysis, if the operating signal bandwidth of both networks is BW, the equivalent noise temperature of the ith two-port active network is:

$$T_{\text{ea}i} = (\text{NF}_i - 1)T_0 \tag{24}$$

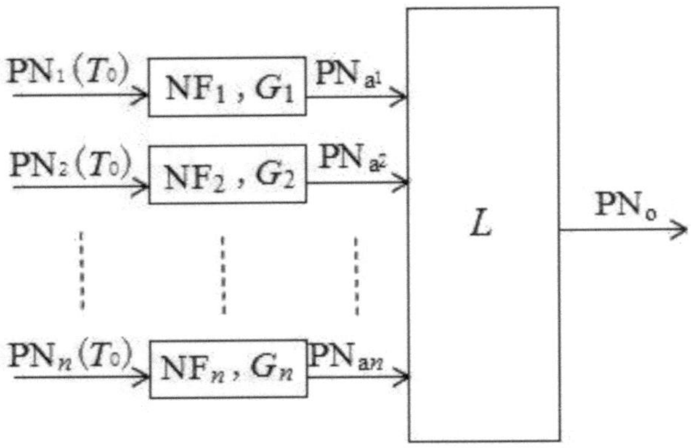

Figure 5.
Cascade of n two-port active networks and multi-port passive networks.

The noise power of the ith two-port active network at its output is:

$$\mathrm{PN}_{\mathrm{a}i} = kT_0\mathrm{BW}G_i + kT_{\mathrm{ea}i}\mathrm{BW}G_i \tag{25}$$

The noise power generated by the ith two-port active network at the output of the passive network is:

$$\mathrm{PN}_{\mathrm{oa}i} = \frac{kT_0\mathrm{BW}G_i + kT_{\mathrm{ea}i}\mathrm{BW}G_i}{nL} \tag{26}$$

The total noise power at the output of the synthesis network is obtained from Eqs. (16) and (26):

$$\begin{aligned}\mathrm{PN}_{\mathrm{o}} &= \mathrm{PN}_{\mathrm{oL}} + \sum_{i=1}^{n}\mathrm{PN}_{\mathrm{oa}i} = \left(1-\frac{1}{L}\right)kT_0\mathrm{BW} + \sum_{i=1}^{n}\frac{kT_0\mathrm{BW}G_i + kT_{\mathrm{ea}i}\mathrm{BW}G_i}{nL}\\ &= \left(1-\frac{1}{L}\right)kT_0\mathrm{BW} + \left[\frac{k\mathrm{BW}}{nL}\sum_{i=1}^{n}(T_0+T_{\mathrm{ea}i})G_i\right]\end{aligned} \tag{27}$$

We can calculate the total noise figure NF of the synthetic network as follows:

$$\begin{aligned}\mathrm{NF} &= \frac{\mathrm{PN}_{\mathrm{o}}}{kT_0\mathrm{BW}G_{\Sigma}} = \mathrm{PN}_{\mathrm{o}}\Big/\left(kT_0\mathrm{BW}\sum_{i=1}^{n}\frac{G_i}{nL}\right)\\ &= \left\{\left(1-\frac{1}{L}\right)kT_0\mathrm{BW} + \left[\frac{k\mathrm{BW}}{nL}\sum_{i=1}^{n}(T_0+T_{\mathrm{ea}i})G_i\right]\right\}\Big/\left(kT_0\mathrm{BW}\sum_{i=1}^{n}\frac{G_i}{nL}\right)\\ &= n(L-1)\Big/\sum_{i=1}^{n}G_i + \left[\sum_{i=1}^{n}\left(1+\frac{T_{\mathrm{ea}i}}{T_0}\right)G_i\right]\Big/\sum_{i=1}^{n}G_i\end{aligned} \tag{28}$$

where $G_{\sum}$ is the gain of the synthetic network.

If the gain and noise figure of the two-port active network are the same, i.e., $G_i = G$ and $\mathrm{NF}_i = \mathrm{NF}_1$, then Eq. (28) is simplified as:

$$\mathrm{NF} = 1 + \frac{T_{\mathrm{ea}i}}{T_0} + \frac{L-1}{G} = \mathrm{NF}_1 + \frac{L-1}{G} \tag{29}$$

6.1 Analysis of noise characteristics of active and passive synthetic networks with $n-1$ input ports connected to a matching load with a noise temperature of T_0

The previous analysis is to analyze the noise characteristics of n two-port active networks under normal operation. If the $n-1$ input ports of the synthesis network are connected to a matching load with a noise temperature of T_0 and all two-port active networks operate normally, what will happen to the noise characteristics of the synthesis network? At this time, the multi-port synthesis network degenerates into a two-port network, but the total noise power at its output remains unchanged, the same as Eq. (27). At this time, the total noise figure NF of the synthesis network is:

$$\begin{aligned} \mathrm{NF} &= \frac{\mathrm{PN_o}}{kT_0\mathrm{BW}G_\Sigma} = \mathrm{PN_o}\Big/\left(\frac{kT_0\mathrm{BW}G_i}{nL}\right) \\ &= \left\{\left(1-\frac{1}{L}\right)kT_0\mathrm{BW} + \left[\frac{k\mathrm{BW}}{nL}\sum_{i=1}^{n}(T_0+T_{\mathrm{ea}i})G_i\right]\right\}\Big/\left(\frac{kT_0\mathrm{BW}G_i}{nL}\right) \\ &= n(L-1)/G_i + \left[\sum_{i=1}^{n}\left(1+\frac{T_{\mathrm{ea}i}}{T_0}\right)G_i\right]\Big/G_i \end{aligned} \tag{30}$$

If the gain and noise figure of the two-port active network are the same ($G_i = G$, $\mathrm{NF}_i = \mathrm{NF}_1$), the above equation becomes:

$$\mathrm{NF} = n\left(1+\frac{T_{\mathrm{ea}i}}{T_0}+\frac{L-1}{G}\right) = n\left(\mathrm{NF}_1+\frac{L-1}{G}\right) \tag{31}$$

6.2 It is stated in Section 6.1 that if $n-1$ two-port active networks are not operating, the noise characteristics of the synthetic network are analyzed

Following the previous analysis, when the $n-1$ two-port active network input ports in the synthetic network are connected to a matching load with a noise temperature of T_0 and the $n-1$ active networks do not operating, what will happen to the noise characteristics of the synthetic network at this time? In order to facilitate the analysis, it is assumed that the two-port active network matches the passive network in both operating and nonoperating states. At this time, the synthetic network degenerates into a two-port network with nL loss. As mentioned in Section 5, the noise power of the two-port active network at the output port of the synthetic network is:

$$\mathrm{PN_{oa}} = \frac{kT_0\mathrm{BW}G}{nL} + \frac{kT_{\mathrm{ea}}\mathrm{BW}G}{nL} \tag{32}$$

As mentioned in Section 4, the noise power of the passive network at the output port of the synthetic network is:

$$\mathrm{PN_{oL}} = \frac{nk(L-1)T_0\mathrm{BW}}{nL} + \frac{(n-1)kT_0\mathrm{BW}}{nL} = kT_0\mathrm{BW}\left(1-\frac{1}{nL}\right) \tag{33}$$

Then the total noise power at that output port of the synthetic network is:

$$\mathrm{PN_o} = \mathrm{PN_{oa}} + \mathrm{PN_{oL}} = \frac{k(T_0+T_{\mathrm{ea}})\mathrm{BW}G}{nL} + \left(1-\frac{1}{nL}\right)kT_0\mathrm{BW} \tag{34}$$

The noise figure NF of the synthetic network is:

$$\mathrm{NF}=\frac{\mathrm{PN_o}}{kT_0\mathrm{BW}G_\Sigma}=\ \mathrm{PN_o}/\left(\frac{kT_0\mathrm{BW}G}{nL}\right)=\left(1+\frac{T_{eai}}{T_0}+\frac{nL-1}{G}\right)=\mathrm{NF}_1+\frac{nL-1}{G} \tag{35}$$

when n = 1, the synthetic network degrades to section 5 state, that is, the cascade of two-port active network and two-port passive network.

7. Design, application, and verification

7.1 Calculation of array receiving dynamics of phased array radar

An active phased array radar is composed of 64 identical T/R modules and a 64:1 multi-port passive in-phase power synthesis network. Its structure is similar to that of **Figure** 5. In the engineering design, the design specification of the gain and noise figure of all T/R modules are the same, so we use the same noise figure NF_1 and gain G in the analysis and calculation. The error caused by the inconsistent indexes of different T/R modules is always acceptable and reasonable in the engineering design and calculation. Then the total noise power of the synthetic output received by the radar array can be calculated from Eq. (27):

$$\mathrm{PN_o}=kT_0\mathrm{BW}\left(\frac{\mathrm{NF}_1G}{L}+1-\frac{1}{L}\right) \tag{36}$$

Using Eq. (29), NF_1 is expressed by the total noise figure NF of the synthesis network and substituted into the simplified equation above to obtain:

$$\mathrm{PN_o}=kT_0\mathrm{BW}\frac{\mathrm{NF}G}{L} \tag{37}$$

Assuming that the baseband signal bandwidth of the receiver is 4 MHz, the noise figure of the T/R module is 2 dB, the gain is 25 dB, and the active power loss of the 64:1 power synthesis network is 5 dB, the total output noise power of the synthesis network can be calculated by using Eq. (36) as follows:

$$\mathrm{PN_o}=-114+6+22.02=-85.98\ \mathrm{dBm} \tag{38}$$

In order to facilitate calculation in engineering application, we use T/R module noise figure NF_1 to replace the total noise figure NF of the synthesis network and use Eq. (37) to calculate the total output noise power of the synthesis network, then:

$$\mathrm{PN_o}=-114+6+2+25\text{–}5=-86\ \mathrm{dBm} \tag{39}$$

We compare the calculation results of the above two different methods and find that the difference between them is only 0.02 dB. Therefore, as long as the gain of the active network is much larger than the active power loss of the passive network in engineering application, the error caused by using the noise figure of the active network instead of the noise figure of the synthesis network to calculate the total output noise power of the synthesis network can be ignored, which is enough to meet the requirements of engineering design.

Next, we will calculate the dynamic range of the output signal of the synthesis network.

We assume that the input signal power received by each T/R module of the array is −105 dBm and the phases of the input signals are the same, then the output signal power of the synthesis network is:

$$S_o = -105 + 25\text{–}5 + 18 = -67\ \text{dBm} \tag{40}$$

The input dynamic range of signal power relative noise power (regardless of noise introduced by antenna) is:

$$\text{DR}_i = -105 - (-14 + 6) = 3\ \text{dB} \tag{41}$$

After the signal is synthesized by the network, the output dynamics of the signal power relative noise power is as follows:

$$\text{DR}_o = -67 - (-86) = 19\ \text{dB} \tag{42}$$

Note that when calculating the input and output noise power above, the bandwidth of both must be the same; both are 4 MHz.

Through the calculation of the above practical examples, we can draw a conclusion that when calculating the dynamic range of the network output signals synthesized by the active network and the multi-port passive in-phase network, we must remember that the total noise power output by the network is not added, only the in-phase signals can be added, and the dynamic range of the signal to noise will increase after passing through the synthesized network.

7.2 Method for testing noise figure of synthesis network after cascading two-port active network and multi-port passive network

In the active phased array radar, we design a T/R module, which consists of four identical receiving channels. Finally, the four receiving channels are output through a 4:1 power synthesis network. How to measure the noise figure of the T/R module in practical engineering application? Our common noise figure instruments, such as HP8970B and Agilent N8975A, have only one noise source. At first, engineers measured the noise figure of each receiving channel to be about 8 dB under the condition of normal operation of the four channels. This measurement data is quite different from the actual design specification, and there are obvious problems. Later, when measuring the noise figure of one receiving channel, we turned off the other three receiving channels and measured the noise figure of each channel in turn. At this time, the noise figure of each channel was measured to be about 2 dB, and the result basically met the design requirements.

It is not difficult for us to understand the above phenomena by using the previous analysis and derivation results. Obviously, it can be seen from Eq. (29) that the noise figure of the synthetic network is basically close to that of a single active channel (when G is much larger than nL). Looking at Eq. (31), we find that when all four channels are operating, if we measure the noise figure of one channel, it will increase by n times than the theoretical value, where n is 4, i.e. 6 dB, so the noise figure we measure is about 8 dB. When one receiving channel is measured and the other three receiving channels are turned off, the noise figure measured at this time is the result given by Eq. (35), and the measurement result is close to the noise figure of a single active channel (when G is much larger than nL). Therefore, when

	f_1	f_2	f_3	f_4
Noise figure in single-channel operation/dB	2.28	2.20	2.26	2.24
Noise figure in four-channel operation/dB	8.20	8.12	8.16	8.15

Table 1.
Noise figure of multi-port active synthesis network under different test conditions.

G is much larger than nL (which can be realized in engineering), it can be considered that Eq. (35) is close to Eq. (29). See **Table 1** for noise figure of multi-port active synthesis network under different test conditions.

In engineering applications, we use the existing noise figure test instruments and adopt the above method to measure the noise figure of the multi-port active synthesis network. We must remember that there is a condition that the gain of a single active channel is much larger than the loss of the passive synthesis network (including the distribution loss at this time); otherwise the measurement result will be greatly different from the theoretical value. We can also average the measured values of each channel to characterize the noise figure of the whole synthetic network. For example, the active channel gain G is only 15 dB, while the passive network is 32:1. When the noise figure of the synthesis network is measured by the above test method, the result will cause a large error. The specific reason can be seen in the previous correlation analysis and calculation formula (35). Of course, we can also correct the measurement by setting the loss of the DUT in the noise figure testing instrument, so that we can also obtain the correct measurement value. For specific operation settings, please refer to the relevant operating instructions of the noise figure test instrument.

In this chapter, the mathematical expressions of the total output noise power and noise figure of the multi-port network in many common cases are given. Using these formulas, designers can calculate the dynamic range of the active phased array radar receiving system and can also use the calculation formula of noise figure to formulate the testing scheme of the active phased array radar noise figure [4, 5].

Author details

Yu Hongbiao
Nanjing Research Institute of Electronics Technology (NRIET), China

*Address all correspondence to: sssyhb@126.com

References

[1] Connor FR. Noise. Beijing: Science Press; 1982. p. 72

[2] Sklar B. In: PingPing X et al., editors. Digital Communications Fundamentals and Applications. 2nd ed. Beijing: Electronics Industry Press; 2002. pp. 207-213

[3] Guangyi Z. Phased Array Radar System. Beijing: National Defense Industry Press; 2000. p. 218

[4] Lin GM, Yuan LS. Noise measurement of active phased array receiver system. Modern Radar, China. 2004;**26**(3):54-57

[5] Chuandong H. Measurement of equivalent noise temperature of phased array radar receiving system. In: National Conference on Microwave Measurement. 2000

Chapter 9

Wearable Textile Antennas with High Body-Antenna Isolation: Design, Fabrication, and Characterization Aspects

Nikolay Atanasov, Gabriela Atanasova and Blagovest Atanasov

Abstract

This chapter provides a brief overview of the types of wearable antennas with high body-antenna isolation. The main parameters and characteristics of wearable antennas and their design requirements are discussed. Next, procedures (passive and active) to test the performance of wearable antennas are presented. The electromagnetic properties of the commercially available textiles used as antenna substrates are investigated and summarized here, followed by a more detailed examination of their effects on the performance of wearable antennas with high body-antenna isolation. A trade-off between substrate electromagnetic properties and resonant frequency, bandwidth, radiation efficiency, and maximum gain is presented. Finally, a case study is presented with detailed analyses and investigations of the low-profile all-textile wearable antennas with high body-antenna isolation and low SAR. Their interaction with a semisolid homogeneous human body phantom is discussed. The simulations and experimental results from different (in free-space and on-body) scenarios are presented.

Keywords: wearable antenna, flexible antenna, textile antenna, design requirements, SAR, antenna measurements, antenna performance, human body phantom

1. Introduction

The wearables are identified as 1 of the 10 technologies which will change our lives [1]. They offer attractive solutions in diverse areas including healthcare, education, finance, sport, and entertainment. For example, in the area of the healthcare, wearable devices can collect data (on blood pressure, temperature, heart rate, steps, calories burned, and even glucose levels) in real-time and send this information to nearby node (on-body communication between two wearable devices) or remote station (off-body communication between a wearable device and mobile phone, tablet, or personal computer) using body area networks (BANs). In order to realize remote monitoring and real-time feedback to the user, the wearable device needs to be equipped with a sensor, processor, memory, power unit, transceiver, and an antenna.

The wearable antenna plays a significant role in the overall performance of each wireless wearable device because it determines the reliability of the wireless link and directly influences the energy efficiency and battery life of the device [2, 3]. However, because the wearable antenna operates in a specific environment (on or near to the human body), the effects due to lossy body tissues (as impedance mismatching, radiation-pattern distortion, radiation efficiency reduction) make the design of a wearable antenna a difficult task. Therefore, care is needed in designing antennas for wearable devices [4, 5].

2. Wearable antenna design issues, requirements, measurements, and testing

2.1 Design requirements for wearable antennas

The design of wearable antennas for body-centric communications is discussed in [2–8], and the most important requirements are summarized **Table 1**.

2.2 Design procedure

Generally, in wearable antenna design, electrical, mechanical, and safety requirements should be taken into account. Moreover, to obtain the best antenna performance, the antenna-human interaction needs to be taken into account during the first stage of the wearable antenna design process. As the body is composed of different tissues with different material properties, the choice of a proper body model (called phantom) is critical in ensuring a good trade-off between simulation accuracy and complexity [4]. For the initial design, the simplest and fastest option is to use a homogenous flat phantom of the human body. A detailed review of various types of human body models is presented in [3, 11, 12].

Requirements	
Electrical	• Bandwidth: sufficient to cover the frequency bands over which the wearable device is intended to operate ($\lvert S11\rvert \leq -10$ dB within the frequency band of interest in both on-body and free-space operation conditions) • High efficiency: more than 50% in both on-body and free-space operation conditions • Appropriate radiation patterns that reduce the electromagnetic radiation toward the human body
Mechanical	• Low-profile • Lightweight • Compactness • Flexibility • Robust
Safety	• Low specific absorption rates (SAR): below worldwide standard limits (1 g-SAR < 1.6 W/kg [9] and 10 g-SAR < 2 W/kg [10])
Manufacturing	• Low cost in both materials and fabrication • Simple structure for ease fabrication and mass production

Table 1.
General design requirements for wearable antennas.

The selection of materials for the conductive and non-conductive elements of the antenna is also an important factor to consider, especially when the antenna is required to possess characteristics such as low-profile, lightweight, compactness, flexibility, and robust. Hence, flexible, thinner, and low-cost materials should be chosen to make the antenna conformable to the person wearing the wearable device and to meet mechanical and manufacturing requirements. Materials like polymers [3, 13], non-conductive fabrics [4], paper [14], and flex film [5, 6] have been used as the substrates in the existing wearable antennas [3, 15]. The choice of material for the antenna's substrate is a critical factor in the performance of the antenna and is examined in depth in Section 3. After the selection of a suitable material for the antenna substrate, its electromagnetic properties (complex permittivity and permeability and loss tangent) must be characterized via measurements [4]. Several (resonant and non-resonant) methods described in [16] can be used for characterization of the electromagnetic properties of flexible materials. For conductive antenna elements, thinner copper or brass foils, electrically conductive fabrics, threads, or ink can be chosen.

Finally, the impact of wearable antennas on the human body also needs to be considered. To study possible effects on body tissues, we must examine the rate at which energy (W) is deposited in a given volume (V) of tissue with specific density (ρ), as shown in Eq. (1):

$$SAR = \frac{d}{dt}\left(\frac{dW}{\rho dV}\right)\left[\frac{W}{kg}\right] \tag{1}$$

SAR can be also calculated from the electric field (E) within the tissue, as shown in Eq. (2):

$$SAR = \frac{\sigma E^2}{\rho}\left[\frac{W}{kg}\right], \tag{2}$$

where σ is the electrical conductivity of the tissue [S/m].

To control the possibility of high local peaks, the maximum permitted SAR is specified as applying to any 1 g or 10 g of tissue [17].

Therefore, the antenna topology with high body-antenna isolation is required to guarantee satisfactory performance and to reduce the SAR when the antenna is placed on the human body. Several antenna designs with a high degree of isolation between the antenna and human tissues have been reported. These designs use a full ground plane [15, 18], an artificial magnetic conducting surface [19], a reflector [13], an electromagnetic bandgap structure [20], or substrate integrated waveguide techniques [2, 21].

Based on the above requirements, a flexible wearable antenna with a low profile, high radiation efficiency, and low SAR can be developed using the algorithm for numerical design and optimization proposed in [3].

After that, a prototype of the optimized design of the wearable antenna can be fabricated using the methods for fabrication of flexible and wearable antennas presented in [5, 8, 13].

Finally, the antenna designs need to be confirmed by both numerical simulations and experimental measurements, first in free space and after that, when antennas are placed on a human body model.

2.3 Measurements and testing

During the wearable antenna design and development process, measurements of classical parameters that describe the antenna's performance such as reflection coefficient magnitude (|S11|), bandwidth, gain, radiation efficiency, and radiation patterns need to be performed using passive approaches. In passive antenna measurements, the prototype is connected to the measuring equipment (a network analyzer, signal generator, receiver, or spectrum analyzer) using an external coaxial cable. Moreover, full verification of the antenna design requires more extensive testing, such as flexibility tests (described in [4, 5]) and tests which represent the behavior of the antenna in real working conditions (also called active antenna measurements). In these measurements, a wearable device simulator (or a radio communication test module) is used to set up a connection to the antenna under test, which is embedded into a complete operating wearable device (or connected to a radio communication module) to reproduce real-world behavior. In order to conduct accurate and repeatable measurements, a test (anechoic or reverberation) chamber with a controlled environment is required. The schematic setup for passive (cable-fed) and active testing of antennas in an anechoic and reverberation chamber can be found in [22].

Moreover, different measurement scenarios should be investigated to guarantee optimal antenna performance in a variety of operating conditions: free-space (the antenna is in an isolated test fixture made from foam, placed away from the human body) and on-body scenarios (the antenna is very close to the human body phantom or in direct contact with the phantom).

For wearable applications, the effect of antennas on the human body also must be quantified [8]. The SAR distribution can be measured by the thermographic method (described in [23]) or by a commercial DASY-4 system (presented in [24]).

3. The effect of electromagnetic properties of the substrate materials on the performance of wearable antennas with high body-antenna isolation

The major challenge of designing wearable antennas is to make an antenna that is invisible and unobtrusively integrated inside a garment as well as comfortable and non-hindering for the wearer [7, 25]. The integration of antennas into clothes involves using textile materials as dielectric substrates [7].

The substrate material offers not only ergonomic properties and ease of integration into the users' garments, but it impacts on the antenna performance. Moreover, the thickness of the substrate also influences the overall antenna dimensions [26]. A brief survey on electromagnetic (EM) properties of textile materials used in wearable antennas can be found in [4, 26, 27]. However, little information can be found on the effect of EM properties of the substrate materials on the performance of wearable antennas with body-antenna isolation. The following subsections describe the characterization of the EM properties of textile materials (polar fleece, polyester, polyamide-elastane, cotton, and denim) and their effects on performance (resonant frequency, bandwidth, radiation efficiency, and maximum gain) of wearable antennas with body-antenna isolation.

3.1 Characterization of electromagnetic properties of the textile materials

The EM properties (real (ε') and imaginary (ε'') parts of the relative permittivity and loss tangent ($tan\delta$)) of the textile materials were measured by the cavity perturbation method using a rectangular cavity resonator. This method was chosen because it is well known for extracting EM properties of dielectrics, semiconductors, and magnetic and composite materials [16]. All measurements were performed at 2.564 GHz, at temperature 24 ± 1°C. For each textile material, five readings were

			EM properties			
	Thickness mm	Material layers	ε'	ε''	$tan\delta$	Density g/cm^3
Polar fleece	1.5	One	1.21831	0.00221	0.00183	0.20
Polyester	0.35	One	1.49797	0.00578	0.00389	1.38
	1.5	Four	1.62022	0.00824	0.00509	
Polyamide-elastane	0.5	One	1.52389	0.03103	0.02040	1.14
	1.5	Three	1.54927	0.02268	0.01463	
Cotton	0.52	One	1.63850	0.10199	0.06218	1.52
	1.5	Three	1.63215	0.08049	0.04930	
Denim	0.5	One	1.86986	0.11786	0.06310	1.54
	1.5	Three	1.87813	0.11166	0.05942	

Table 2.
Parameters of the textile materials and results from measurements of their EM properties.

taken, and the average value is given in **Table 2**. To see how material thickness affects the EM properties, we measured ε′ and ε″ of the materials at two thicknesses. The first one corresponds to the thickness of one layer of each material. The second was fixed (1.5 mm) for all textile materials—in this case, three/four layers of each material were assembled by an iron-on hemming strip added in between the martial layers to form a 1.5-mm-thick sample.

From the results presented in **Table 2**, it can be observed that the EM properties of the textile materials show a variation with increasing number of layers in the sample. Moreover, the ε′ shows small variations with respect to the sample thickness in comparison to ε″ and $tan\delta$. For all materials are observed that with increasing of the thickness of the textile sample, the real part of relative permittivity increases. The ε′ value was found to vary between 1.50 (one layer) and 1.62 (four layers) for polyester and between 1.87 (one layer) to 1.88 (three layers) for denim fabric. Another interesting behavior observed is that ε″ and $tan\delta$ of the polyester increase with increasing number of layers (thickness) of the textile sample, while ε″ and $tan\delta$ of the polyamide-elastane, cotton, and denim decrease. These increases are attributed to the presence of glue (in iron-on hemming strip added in between the martial layers), having a high $tan\delta$ compared to the polyester. Indeed, the use of an interface to assemble the layers, such as glue or an adhesive sheet, can lead to variation in EM properties of the textile materials used in the wearable antenna design.

A comparison of different textile materials at the same thickness (1.5 mm) shows that denim has the highest values of ε′, ε″, and $tan\delta$. Moreover, the denim and cotton show higher values of $tan\delta$ than polar fleece, polyester, and polyamide-elastane. The higher values of $tan\delta$ can be related to the fact that these textiles are built up from natural fibers. We conclude that the fabrics made from synthetic fibers (polar fleece, polyester, and polyamide-elastane) have a lower relative permittivity and $tan\delta$ than the fabrics made from natural fibers (cotton and denim).

3.2 Effect of textile substrate electromagnetic property variation on the performance of wearable antennas with body-antenna isolation

Based on the EM properties provided by the above subsection, five textile wearable antennas with high body-antenna isolation were designed using the xFDTD (xFDTD, Remcom Inc., State College, PA, USA), a finite-difference time-domain

(FDTD) method-based simulation software. The configuration of the wearable textile antenna with a substrate from denim fabric is illustrated in **Figure 1a**. It consists of a hexagonal shaped monopole on the top of the substrate and a planar rectangular reflector on the bottom. The antenna is fed by a coplanar waveguide (CPW) feed line. This antenna structure is chosen due to its advantages of light-weight, low-profile, low-cost, and easy fabrication, which satisfies the requirements for wearable antennas presented in **Table 1**.

Then, five antennas were manufactured by a cost-effective and time-saving fabrication technique, as described in [28]. The radiating elements of the antennas were built using a highly conductive woven fabric P1168 (supplied by Adafruit, Italy) with a thickness of 0.08 mm and sheet resistance of 0.05 Ω/sq. **Figure 1b–f** shows the fabricated prototypes of the antennas.

Two scenarios were investigated in this subsection to study the effect of the EM properties of the textile substrate on the antenna performance.

At the first stage, a wearable antenna with a substrate from polar fleece was designed (**Figure 2**) to operate in the 2.4–2.48 GHz industrial scientific and medical (ISM) band. The geometrical dimensions of the monopole, CPW, and reflector

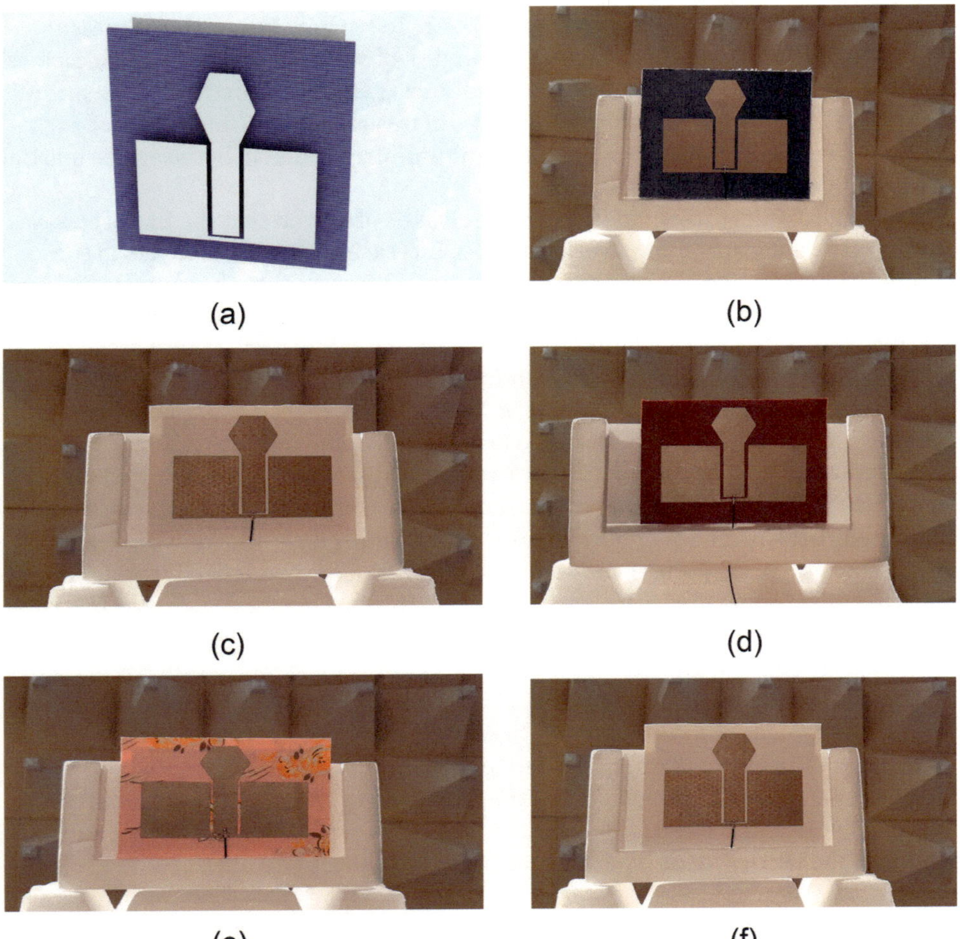

Figure 1.
The (a) 3D numerical model of the textile antenna with a substrate from denim fabric and photographs of the fabricated prototypes, (b) antenna with a denim substrate, (c) antenna with a cotton substrate, (d) antenna with a polyester substrate, (e) antenna with a polyamide-elastane substrate, and (f) antenna with a polar fleece substrate.

were tuned by numerical simulations, following the optimization procedure in [3], to achieve the optimal impedance match, high radiation efficiency, and high front-to-back (FB) ratio at the targeted ISM band. The dimensions of the antenna are listed in **Table 3**.

For comparison purposes, five different numerical models of the wearable antenna with different substrates (from cotton, denim, polyester, polyamide-elastane, air) and geometrical dimensions as that of the antenna with a substrate from polar fleece were built. The resonant frequency of the antennas calculated by the simulations is compared in **Figure 3a**.

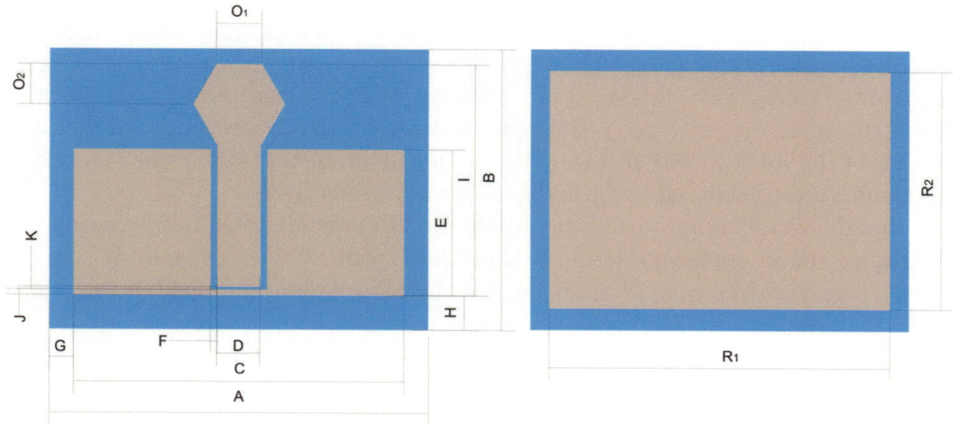

Figure 2.
Configuration of the wearable antenna.

	Antenna with an air-filled substrate	Antenna with a polyester substrate	Antenna with a polyamide-elastane substrate	Antenna with a cotton substrate	Antenna with a denim substrate	Antenna with a polar fleece substrate
A	141.0	141.0	141.0	141.0	141.0	141.0
B	103.5	103.5	103.5	103.5	103.5	103.5
C	134.0	110.0	112.0	109.0	103.0	123.0
D	21.0	21.0	21.0	21.0	21.0	21.0
E	59.0	47.0	48.0	46.5	43.5	53.5
F	2.5	2.5	2.5	2.5	2.5	2.5
G	—	16.5	15.5	17.0	20.0	10.0
H	—	16.5	15.5	17.0	20.0	10.0
I	90.0	78.0	79.0	77.5	74.5	84.5
J	2.0	2.0	2.0	2.0	2.0	2.0
K	1.0	1.0	1.0	1.0	1.0	1.0
O_1	17.0	17.0	17.0	17.0	17.0	17.0
O_2	17.0	17.0	17.0	17.0	17.0	17.0
R_1	138.0	127.0	127.0	127.0	127.0	127.0
R_2	93.0	87.5	87.5	87.5	87.5	87.5

Table 3.
The geometrical dimensions of the optimized antennas.

As expected, the antenna with a substrate filled by air ($\varepsilon_r' = 1$) shows the highest value of resonant frequency (f_{r0}) of 2.66 GHz. For antennas with substrates from textile materials, the resonant frequency drops (to a value of $f_{r0}/\sqrt{\varepsilon_r'}$) as the ε_r' increase, because more energy is coupled into guided waves inside the substrate.

Next, the effects of the EM properties of the textile substrate on the bandwidth were evaluated and illustrated in **Figure 3b–d**. The bandwidth of all antennas was defined for |S11| = −10 dB. **Figure 3b** shows the bandwidth as a function of the ε_r' of the textile substrates. It is observed that the ε_r' of the textile substrate does not strongly influence the bandwidth. This effect can be attributed to the fact that ε_r' of the textiles are within the range between 1.2 and 1.9 (see **Table 2**). Moreover, the bandwidth increases with an increase in the ε_r'' (see **Figure 3c**).

As shown in **Figure 3d**, the antennas with a substrate from denim and cotton have the largest bandwidth (150.5 MHz) because these two fabrics have the highest values of ε_r'' and $tan\delta$ at a thickness of 1.5 mm (see **Table 2**) and, thus, the highest dielectric losses. As expected, a decline in the bandwidth is observed as the dielectric loss of the substrate is decreased. Hence, it can be concluded that the antenna bandwidth is proportional to the textile substrate loss tangent.

From the results mentioned here, it also can be concluded that fabrics made from synthetic fibers (polar fleece, polyamide-elastane, polyester) exhibit a narrow bandwidth compared to the fabrics made from natural fibers (cotton and denim).

Moreover, it is well known that the bandwidth and radiation efficiency not only are determined by the substrate's EM properties and thickness but

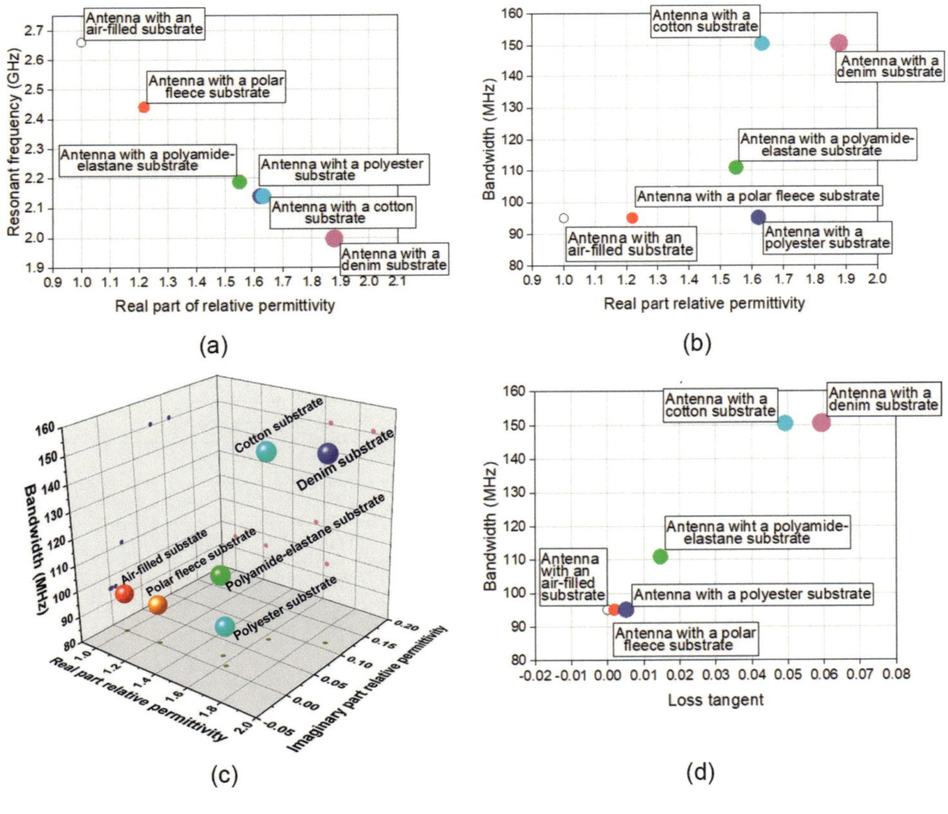

Figure 3.
Variation of (a) resonant frequencies and (b) bandwidths with the ε_r', (c) bandwidths with ε_r' and ε_r'', and (d) bandwidths with the loss tangent of the textile substrates used in wearable antennas with the same geometrical dimensions.

also depend on the size of the antenna radiating elements and matching. Consequently, it is necessary to know the effect of the EM properties of the textile materials on antenna performance when antenna elements are optimized. For this reason, in the second scenario, the structure of the antennas with a polyester, polyamide-elastane, denim, cotton, and air-filled substrate was optimized for the maximum impedance bandwidth, radiation efficiency, and high FB ratio at the 2.4–2.48 GHz frequency band. The optimized dimensions are listed in **Table 3**.

To illustrate the effects of textile materials on the antenna performance, the optimized antenna designs considered in **Table 3** were evaluated in the free space and when placed on a flat phantom. First, a flat homogeneous semisolid phantom of the human body with dimensions 265 × 50 × 350 mm to emulate 2/3 muscle tissue was fabricated accordingly to the recipe and technique described in [29]. After the phantom mixture had solidified, EM properties were measured at 2.564 GHz: $\varepsilon_r' = 40.805$ and $\sigma = 2.33$ S/m. Next, a numerical flat homogeneous human body model (called flat phantom) with dimensions and EM properties as those of the fabricated phantom was developed. The fabricated experimental and numerical phantoms used in this chapter are illustrated in **Figure 4**.

The results from numerical simulations in both free space and on-body are presented in **Figures 5–9**. **Figure 5** displays the bandwidth as a function of the loss tangent of the textile substrate. As seen in these plots, the bandwidth remains unchanged in both free space (**Figure 5a**) and on-body (**Figure 5b**) for the antennas with a polyester, cotton, and denim substrate. A slight bandwidth broadening is observed when the optimized antennas with an air-filled, polar fleece and polyamide-elastane substrate are placed on the flat phantom. The reason for this effect

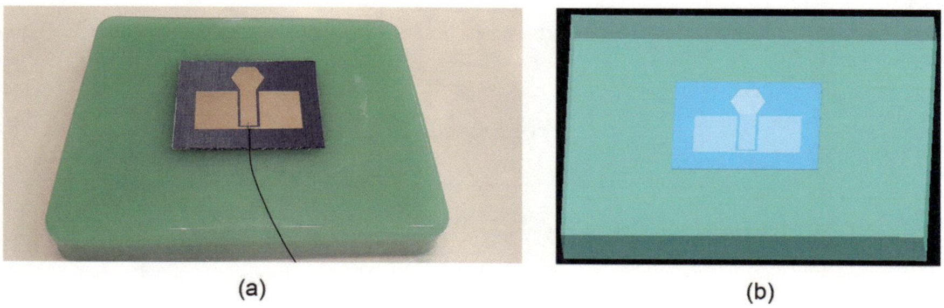

Figure 4.
Flat phantom and the antenna with a denim substrate: (a) a photograph and (b) numerical models.

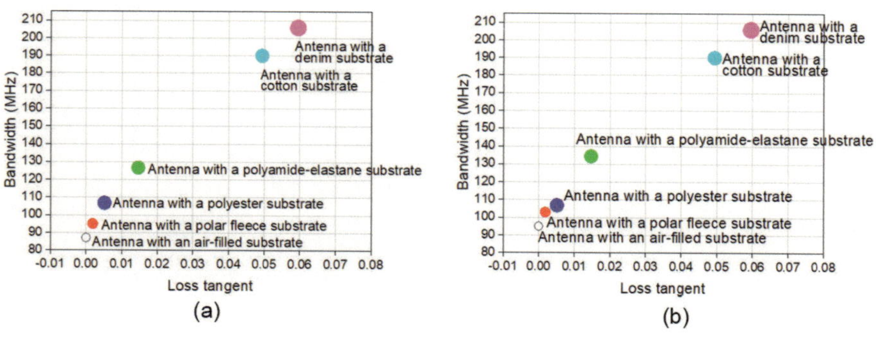

Figure 5.
Variation of the bandwidth with the loss tangent of the textile substrates used in optimized wearable antennas with geometrical dimensions listed in **Table 3**, *in both (a) free space and (b) on-body.*

is that the real part of the relative permittivity of these materials (at a thickness of 1.5 mm, see **Table 2**) is between 1 and 1.55. Consequently, we can conclude that the performance of the wearable antennas with substrates made from fabrics with real part of the relative permittivity small than 1.6 will be influent from the proximity of the human body.

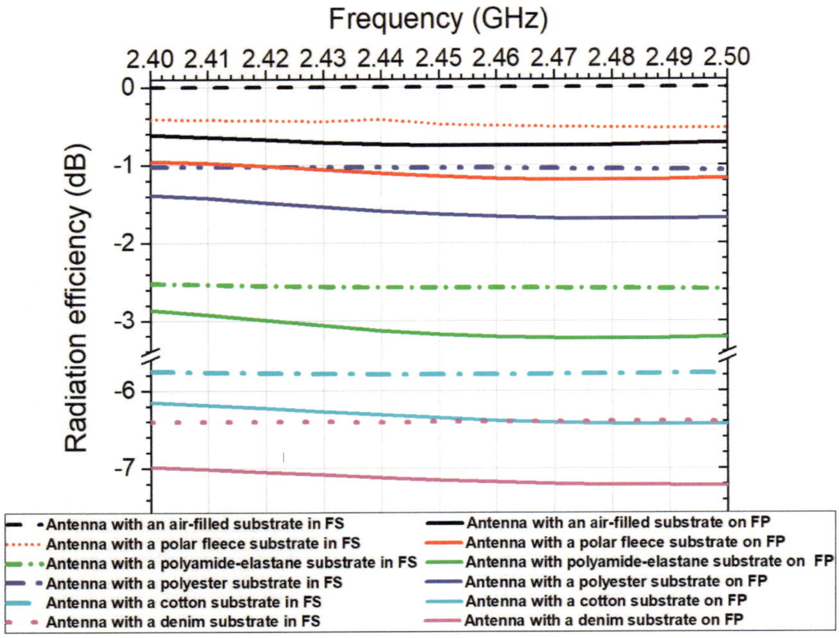

Figure 6.
Simulated radiation efficiency curves versus frequency of the wearable antennas: in the free space (FS) and on the flat phantom (FP).

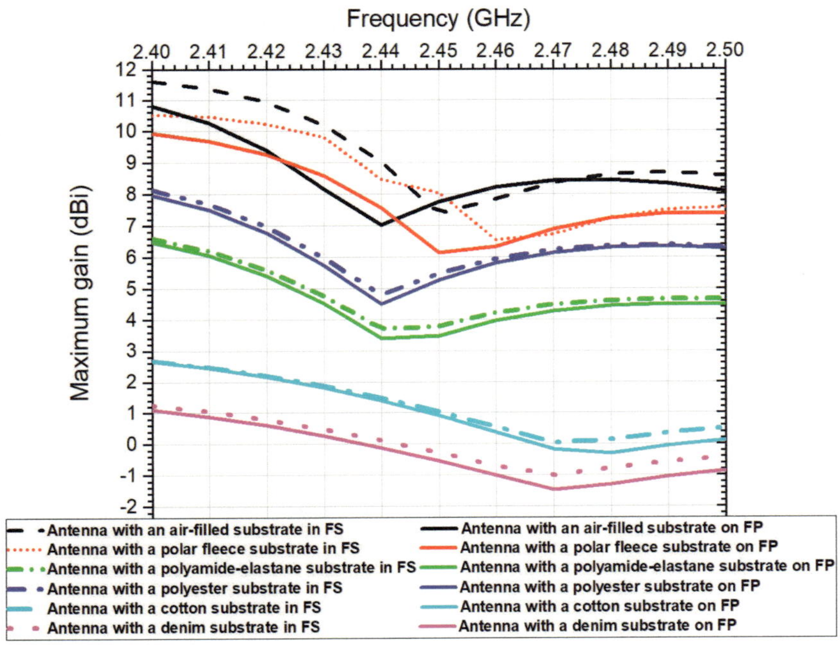

Figure 7.
Simulated maximum gain curves versus frequency of the wearable antennas: in the free space (FS) and on the flat phantom (FP).

DOI: http://dx.doi.org/10.5772/intechopen.91143

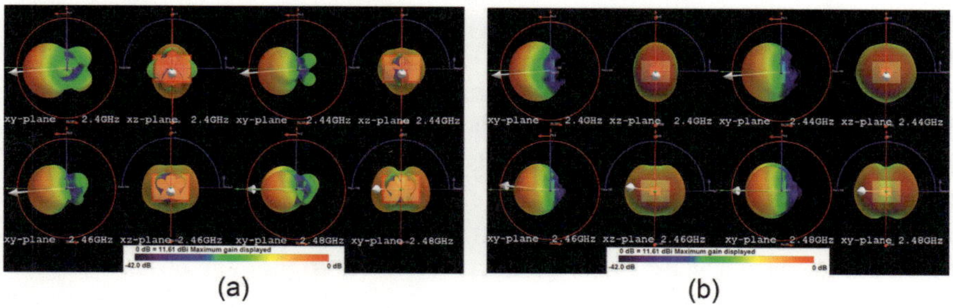

Figure 8.
Simulated three-dimensional radiation patterns at 2.4, 2.44, 2.46, and 2.48 GHz for the antenna with a polar fleece substrate (a) in the free space and (b) on the flat phantom.

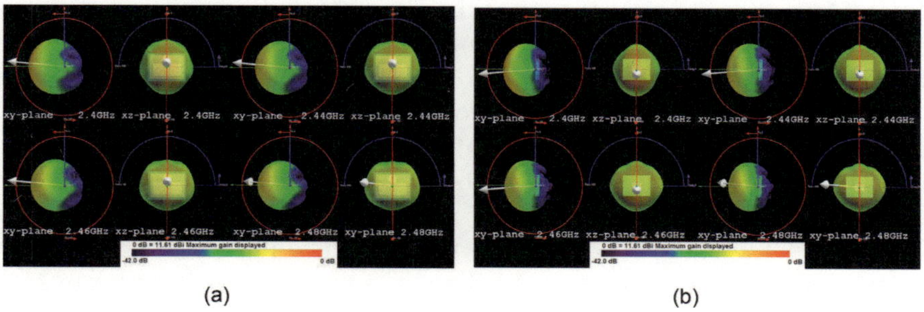

Figure 9.
Simulated three-dimensional radiation patterns at 2.4, 2.44, 2.46, and 2.48 GHz for the antenna with a denim substrate (a) in the free space and (b) on the flat phantom.

A comparison between the simulated radiation efficiency of the antennas is displayed in **Figure 6**. The radiation efficiency is defined as the ratio of the power radiated from the antenna to the net input power, which is the radiated power plus material losses [30]. The comparison shows that the antennas with polar fleece, polyester, and polyamide-elastane substrates achieve a much better radiation efficiency than the antennas with substrates from cotton and denim. These differences are attributed to the fact that fabrics made from synthetic fibers have lower $tan\delta$ than the fabrics made from natural fibers. Moreover, it can be observed that the simulated radiation efficiency of all antennas is quite stable in the target frequency range, on both the free space and on-body.

From the results presented in **Figure 6**, a slight reduction of radiation efficiency when the antennas are placed on the flat phantom (FP) also can be observed. For example, across the operating band, the radiation efficiency of the antenna with a substrate from polar fleece is estimated to be −0.47 dB (90%) in the free space and − 1.1 dB (78%) when it is placed on the phantom.

The maximum gain of the optimized antennas was also evaluated and illustrated in **Figure 7**. As seen, in the target frequency band, the gain varies between 11 and 7 dBi (for the antenna with a polar fleece substrate) and between 1 and −1 dBi (for the antenna with a denim substrate). The variation in maximum gain values is related to the maximum directivity (see **Figure 8**) and is primarily due to the coupling between radiating elements and the reflector. The gain difference between the antennas with substrates from fabrics made with synthetic fibers and fabrics made with natural fibers can be associated with the differences in their radiation efficiency (see **Figure 6**). Moreover, the maximum gains of the antennas mounted directly on the flat phantom are not strongly affected by the phantom (human body).

Figures 8 and **9** compare the three-dimensional (3D) radiation patterns of the optimized wearable antennas with polar fleece and denim substrates, at 2.4, 2.44, 2.46, and 2.48 GHz. These frequencies approximately correspond to the lower, middle, and upper end of the 2.45 GHz ISM band. As shown in **Figures 8** and **9**, the radiation is unidirectional for the antennas at all frequencies. Three-dimensional patterns show that a small amount of the energy is radiated in the backward direction (i.e., behind the antenna) into the human body. Moreover, it can be observed that in the free space, the Eφ is the dominant field component at 2.4, 2.44, and 2.46 GHz for the antennas. At 2.48 GHz, the Eθ is the dominant field component for these antennas. Hence, we observe that the directivity is decreased with increasing frequency because more energy is radiated in θ direction.

As can be seen in the figures, the radiation patterns of the antennas are not significantly modified by the presence of the flat phantom.

Figure 10 shows FB radio as a function of frequency both in free space and on a phantom. As seen, in the target frequency band, the FB ratios vary between 12 and 33 dB in the free space and between 18 and 36 dB on the phantom, depending on the antenna's substrate. A FB ratio of about 30 dB is achieved for the antennas with denim and cotton substrates in the free space, in the frequency band of 2.4–2.46 GHz, indicating a small amount of radiation behind the antenna.

As seen in **Figure 10**, both antennas with polyester and polyamide-elastane substrates have FB ratio better than 17 dB in the free space and better than 23 dB on the phantom. Moreover, the antennas with cotton, polyester, polyamide-elastane, and denim preserve their FB ratios when placed directly on the flat phantom. In the case of the antennas with polar fleece and air-filled substrate, we see a FB ratio between 12 and 20 dB in the free space and between 25 and 35 dB when antennas are placed directly on the flat phantom; this will result in an increase in SAR values (see **Figure 11**). Comparing the FB ratios of the antennas, it can be concluded that the cotton and denim textile substrates improve FB ratio of the antenna.

3.3 Effect of wearable antennas with textile substrates on the human body

Because the wearable textile antennas are designed to operate near to the human body, in this subsection, we investigate the effects of these antennas on the human

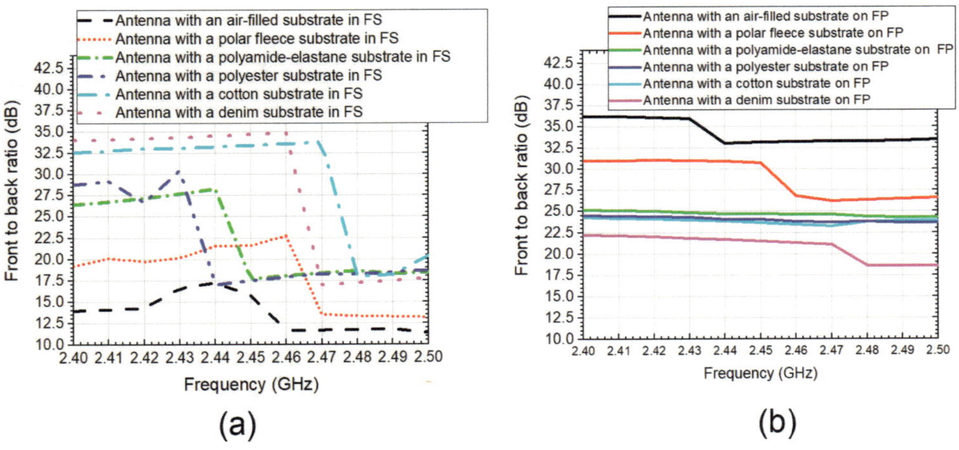

Figure 10.
Simulated front-to-back radio as a function of frequency in both (a) free space and (b) on phantom.

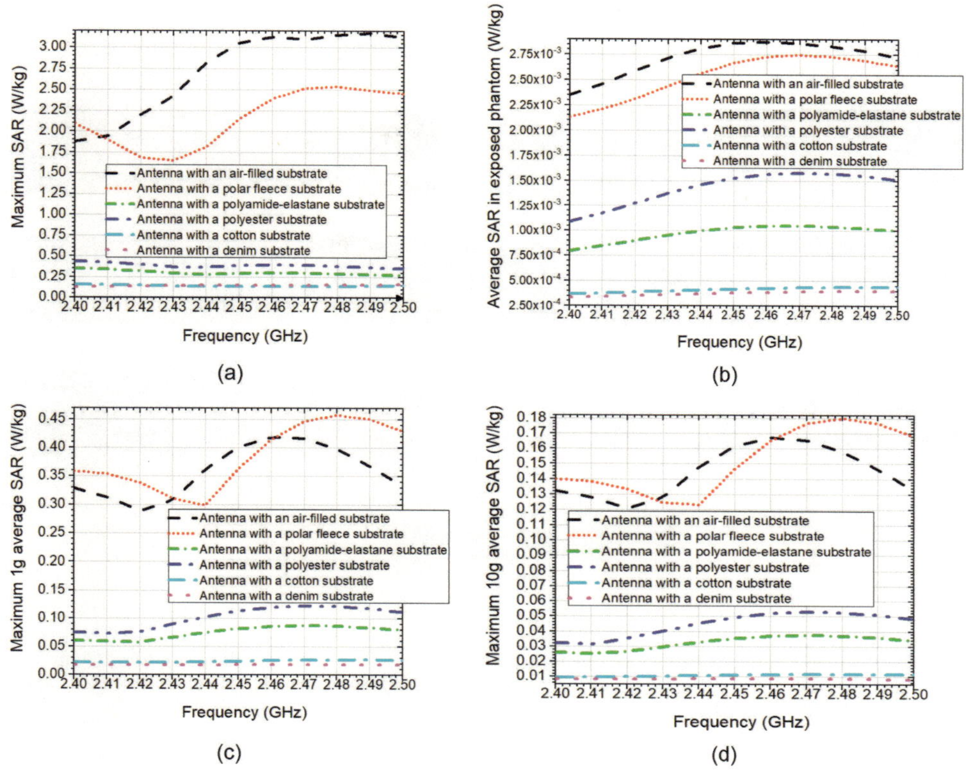

Figure 11.
SAR as a function of frequency (a) the maximum local SAR, (b) average SAR in exposed flat phantom, (c) maximum 1 g average SAR and (d) maximum 10 g average SAR.

body by evaluating the SAR. For these computations, each antenna was placed directly on the numerical flat homogeneous human body model as presented in **Figure 4b**.

Figure 11a–d shows the maximum local SAR, average SAR in exposed flat phantom, and maximum 1 g and 10 g average SAR as a function of frequency. The results presented in **Figure 11a–d** were normalized to net input power of 100 mW. As can be seen in the figures, the SAR values from the antennas with air-filled and polar fleece substrates are higher than the SAR from the other antennas. Moreover, the maximum 10 g average SAR is 0.18 (antenna with a polar fleece substrate) and between 0.04 and 0.1 W/kg for the rest of the antennas. Therefore, maximum 10 g average SAR for all antennas is 90% lower than the specification required by the ICNIRP [10] and also smaller than most of the previously proposed wearable textile antennas.

The differences in SAR distributions between the antennas are illustrated in **Figure 12a–f**. For all antennas, the peak SAR in the phantom occurs in the region near the antenna edges. Moreover, the SAR distribution of the antenna with a polar fleece substrate is similar to that of the antenna with an air-filled substrate. Also, the SAR distribution of the antennas with a substrate from cotton and denim is quite similar.

SAR values presented in **Figure 12** were averaged over a volume of 0.125 mm^3 containing a mass of 0.14575 mg.

Measurements were carried out to validate the simulations. **Figure 13** shows the simulated and measured reflection coefficient magnitudes of the wearable antennas in the free space and on the phantom. A small difference between simulated and measured results was observed. This difference can be attributed to fabrication and assembly inaccuracies (due to the manual assembly).

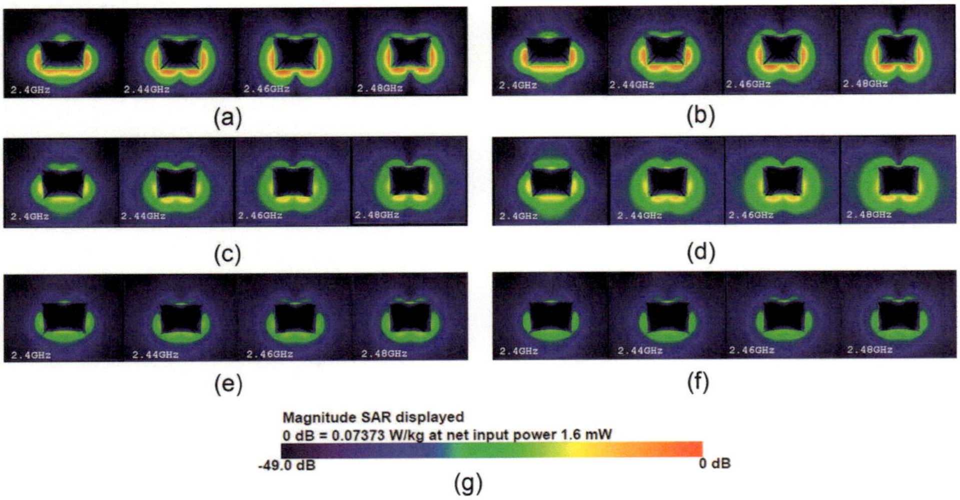

Figure 12.
SAR distributions at 2.4, 2.44, 2.46, and 2.48 GHz for the antenna with (a) an air-filled substrate, (b) a polar fleece substrate, (c) a polyamide-elastane substrate, (d) a polyester substrate, (e) a cotton substrate, (f) a denim substrate, and (g) scale.

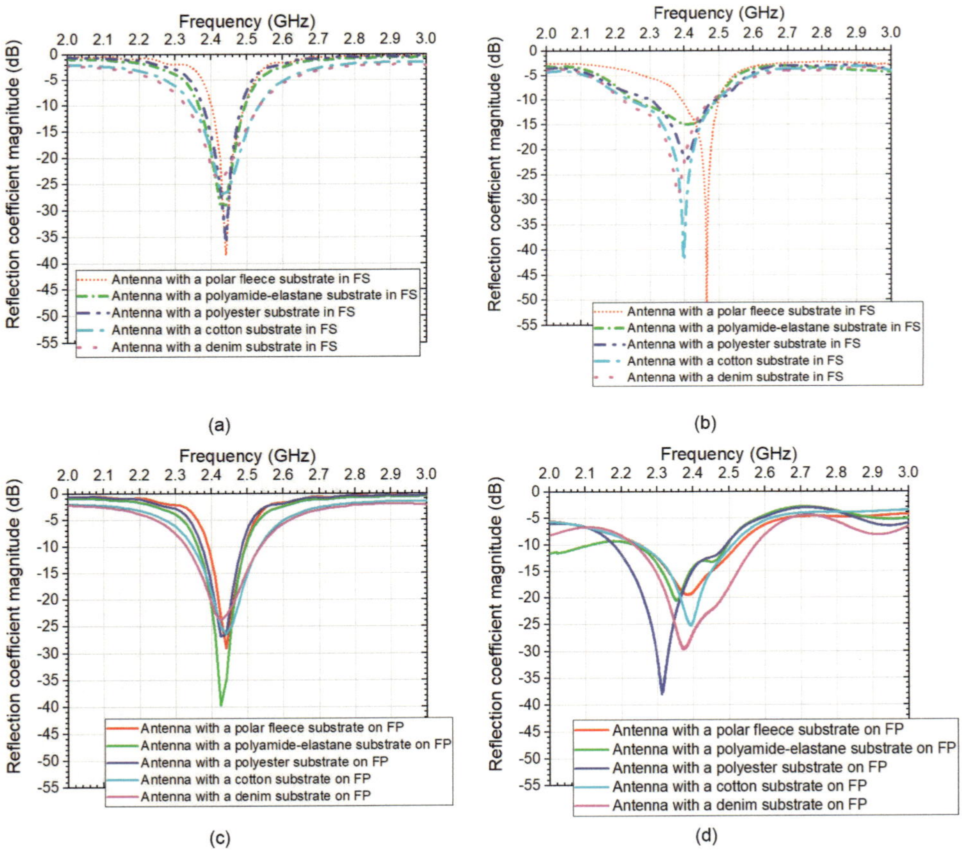

Figure 13.
|S11| curves versus frequency (a) simulated in the free space, (b) measured in the free space, (c) simulated on the phantom, and (d) measured on the phantom.

Moreover, the coaxial cable and U. FL connector were not integrated into the FDTD simulations, which also lead to a difference between the simulations and measurements.

We can conclude that the antennas demonstrate stable performance on both in the free space and when placed directly on the flat phantom.

4. Conclusions

In this chapter, the main parameters and characteristics of wearable antennas and their design requirements have been presented. The electromagnetic properties of the textile substrates also have been examined. From the results mentioned here, it is concluded that fabrics made from synthetic fibers (polar fleece, polyamide-elastane, polyester) have a lower relative permittivity and *tanδ* than the fabrics made from natural fibers (cotton and denim). It is interesting to note that for the wearable antennas with high body-antenna isolation, an increase in the real part of the relative permittivity of the textile substrate had minimal impact on the antenna bandwidth. Moreover, the loss tangent of the textile substrate affects the wearable antenna's bandwidth and radiation efficiency: the higher the loss tangent, the broader the bandwidth and the lower the radiation efficiency. We conclude that the EM properties (ε_r', ε_r'', $tan\delta$), thickness of the textile substrate and radiating element geometry, play an important role in determining the antenna's performance and need to be taken into account when designing a wearable antenna with high body-antenna isolation.

Five low-profile all-textile antennas with high body-antenna isolation have been presented. The in-depth performance evaluations on both in the free space and on the flat phantom of these antennas indicate that they cover 2.45 GHz ISM band and the maximum 10 g average SAR for all antennas is 90% lower than the specification.

Acknowledgements

The author would like to acknowledge the Bulgarian National Science Fund, Ministry of Education and Science, Bulgaria, for the support through a grant № KP-06-H27/11 from 11 December 2018 "Antenna technology for wearable devices in the future communication networks."

Author details

Nikolay Atanasov[1,2]*, Gabriela Atanasova[1] and Blagovest Atanasov[3]

1 South-West University "Neofit Rilski", Blagoevgrad, Bulgaria

2 Electromagnetic Compatibility Laboratory, Bulgarian Institute of Metrology, Sofia, Bulgaria

3 91 German Language High School "Prof. Konstantin Galabov", Sofia, Bulgaria

*Address all correspondence to: natanasov@windowslive.com; natanasov@swu.bg

References

[1] European Commission Directorate-General for Communications Networks, Content and Technology, Smart Wearables: Reflection and Orientation Paper. Digital Industry Competitive Electronics Industry, Brussels; 2016

[2] Agneessens S. Coupled eighth-mode substrate integrated waveguide antenna: Small and wideband with high-body antenna isolation. IEEE Access. 2017;**6**:1595-1602. DOI: 10.1109/ACCESS.2017.2779563

[3] Al-Sehemi A, Al-Ghamdi A, Dishovsky N, Atanasov N, Atanasova G. Wearable antennas for body-centric communications: Design and characterization aspects. Applied Computational Electromagnetics Society Journal. 2019;**34**:1172-1181

[4] Werner D, Jiang Z-H. Electromagnetics of Body-Area Networks, Antennas, Propagations and RF Systems. 1st ed. Hoboken: John Wiley & Sons, Inc.; 2016. pp. 1-59

[5] Khaleel R, Al-Rizzo H, Ayman I. Design, fabrication, and testing of flexible antennas. In: Kischk A, editor. Advancement in Microstrip Antennas with Recent Applications. 1st ed. London: IntechOpen; 2013. pp. 365-383. DOI: 10.5772/50841

[6] Raad H, Abbosh A, Al-Rizzo H, Rucker D. Flexible and compact AMC based antenna for telemedicine applications. IEEE Transactions on Antennas and Propagation. 2013;**61**(2):524-531

[7] Nepa P, Rogier H. Wearable antennas for off-body radio links at VHF and UHF bands: Challenges, state-of-the-art, and future trends below 1 GHz. IEEE Antennas and Propagation Magazine. 2015;**57**(5):30-52. DOI: 10.1109/MAP.2015.2472374

[8] Khaleel H, editor. Innovations in Wearable and Flexible Antennas. 1st ed. Southampton: WIT press; 2015. pp. 8-54. ISBN: 978-1-84-564-986-9

[9] IEEE Standards Coordinating Committee. IEEE for safety levels with respect to human exposure to radio frequency electromagnetic fields, 3 kHz to 300 GHz. IEEE Standards. 2005;**C95**:1

[10] ICNIRP. Guidelines for limiting exposure to time-varying electric, magnetic, and electromagnetic fields (up to 300 GHz). Health Physics. 1998;**74**:494-522

[11] Pellegrini A, Brizzi A, Zhang L, Ali K, Hao Y, Wu X, et al. Antennas and propagation for body-centric wireless communications at millimeter-wave frequencies: A review [wireless corner]. IEEE Antennas and Propagation Magazine. 2013;**55**(4):262-287. DOI: 10.1109/MAP.2013.6645205

[12] Björninen T. Comparison of three body models of different complexities in modelling of equalsized dipole and folded dipole wearable passive UHF RFID tags. Applied Computational Electromagnetics Society Journal. 2018;**33**:706-709

[13] Al-Sehemi A, Al-Ghamdi A, Dishovsky N, Atanasov N, Atanasova G. Design and performance analysis of dual-band wearable compact low-profile antenna for body-centric wireless communications. International Journal of Microwave and Wireless Technologies. 2018;**10**:1175-1185

[14] Anagnostou D, Gheethan A, Amert A, White K. A direct-write printed antenna on paper-based organic substrate for flexible displays and WLAN applications. Journal of Display Technology. 2010;**6**(11):558-564. DOI: 10.1109/JDT.2010.2045474

[15] Simorangkir R, Kiourti A, Esselle K. UWB wearable antenna with full ground plane based on PDMS-embedded conductive fabric. IEEE Antennas and Wireless Propagation Letters. 2018;**17**(3):493-496. DOI: 1109/LAWP.2018.2797251

[16] Chen L, Ong C, Neo C, Varadan VV, Varadan KV. Microwave Electronics: Measurement and Materials Characterization. 1st ed. Chichester: John Wiley & Sons, Inc.; 2004. pp. 37-141

[17] Chen Z-N. Antennas for portable devices. 1st ed. Chichester: John Wiley & Sons, Inc.; 2007. pp. 20-40. ISBN 978-0-470-03073-8

[18] Abbas S, Zahra H, Hashmi R, Esselle K, Volakis J. Compact on-body antennas for wearable communication systems. In: International Workshop on Antenna Technology (iWAT); 3-6 March 2019. Miami, Florida, United Sates of Amerika: IEEE; 2019

[19] Jiang Z-N, Brocker D, Sieber P, Werner D. A compact, low-profile metasurface-enabled antenna for wearable medical body-area network devices. IEEE Transactions on Antennas and Propagation. 2014;**62**(8):4021-4030. DOI: 10.1109/TAP.2014.2327650

[20] Ashyap A, Abidin Z, Dahlan S, Majid H, Kamarudin M, Alomainy A, et al. Highly efficient wearable CPW antenna enabled by EBG-FSS structure for medical body area network applications. IEEE Access. 2018;**6**:77529-77541. DOI: 10.1109/ACCESS.2018.2883379

[21] Agneessens S, Lemey S, Rogier H, Vervust T, Vanfleteren J. Applying QMSIW technique in textile for compact wearable design and high body-antenna isolation. In: IEEE International Symposium an Antennas and Propagation & USNC/URSI Radio Science Meeting; 19-24 July 2015. Vancouver, BC, Canada; 2015

[22] Anguera J, Andújar A, Huynh M-C, Orlenius C, Picher C, Puente C. Advances in antenna technology for wireless handheld devices. International Journal of Antennas and Propagation. 2012;**2013**:1-25

[23] Al-Sehemi A, Al-Ghamdi A, Dishovsky N, Atanasov N, Atanasova G. Flexible and small wearable antenna for wireless body area network applications. Journal of Electromagnetic Waves and Applications. 2017;**31**:1063-1082. DOI: 10.1080/09205071.2017.1336492

[24] Schmid T, Egger O, Kuster N. Automated E-field scanning system for dosimetric assessments. IEEE Transactions on Microwave Theory and Techniques. 1996;**44**(1):105-113. DOI: 10.1109/22.481392

[25] Corchia L, Monti G, Tarricone L. Wearable antennas: Nontextile versus fully textile solutions. IEEE Antennas and Propagation Magazine. 2019;**61**(2):71-83. DOI: 10.1109/MAP.2019.2895665

[26] Salvado R, Loss C, Gonçalves R, Pinho P. Textile materials for the design of wearable antennas: A survey. Sensors. 2012;**12**:15841-15857. DOI: 10.3390/s121115841

[27] Velez F, Miyandoab F, editors. Wearable Technologies and Wireless Body Sensor Networks for Healthcare. 1st ed. London: The Institution of Engineering and Technology; 2019. pp. 147-183. ISBN:978-1-78561-217-6

[28] Atanasov N, Atanasova G, Stefanov A, Nedialkov I. A wearable, low-profile, fractal monopole antenna integrated with reflector for enhancing antenna performance and SAR reduction. In: IEEE MTT-S International

Microwave Workshop Series an Advanced Materials and Processes for RF and THz Applications (IMWS-AMP); 16-18 July 2019. Bochum, Germany: IEEE; 2019. pp. 67-69

[29] Yilmaz T, Foster R, Hao Y. Broadband tissue mimicking phantoms and a patch resonator for evaluating noninvasive monitoring of blood glucose levels. IEEE Transactions on Antennas and Propagation. 2014; **62**(6):3064-3075. DOI: 10.1109/TAP.2014.2313139

[30] Elwi T, Al-Rizzo H, Rucker D, Khaleel H. Effects on twisting and bending on the performance of a miniaturized truncated sinusoidal printed circuit antenna for wearable biomedical telemetry devices. AEU - International Journal of Electronics and Communications. 2010;**65**:217-225. DOI: 10.1016/j.aeue.2010.03.008